Field Guide to the Common Grasses of Oklahoma, Kansas, and Nebraska

Field Guide to the Common Grasses of Oklahoma, Kansas, and Nebraska

Iralee Barnard

 University Press of Kansas

Published by the University Press of Kansas (Lawrence, Kansas 66045), which
was organized by the Kansas Board of Regents and is operated and funded by
Emporia State University, Fort Hays State University, Kansas State University,
Pittsburg State University, the University of Kansas, and Wichita State
University

Publication made possible, in part, by a grant from the Westar Energy Green
Team.

Library of Congress Cataloging-in-Publication Data
Barnard, Iralee.
Field guide to the common grasses of Oklahoma, Kansas, and Nebraska /
Iralee Barnard.
    p.   cm.
Includes bibliographical references and index.
ISBN 978-0-7006-1945-0 (pbk. : alk. paper)   1. Grasses—Oklahoma—
Identification.   2. Grasses—Kansas—Identification.   3. Grasses—Nebraska—
Identification.   I. Title.
SB202.U6B37   2014
633.209766—dc23

                                                              2013035678

British Library Cataloguing-in-Publication Data is available.

Printed in China
10 9 8 7 6 5 4 3 2 1

The paper used in this publication is recycled and contains 30 percent
postconsumer waste. It is acid free and meets the minimum requirements of
the American National Standard for Permanence of Paper for Printed Library
Materials Z39.48-1992.

Photographs: *Cover:* Canada wildrye. *Preceding spread:* Sideoats grama.
Opposite page: January at Kanopolis State Park in central Kansas.

This book is dedicated to

all who love the outdoors and learning.

May your path be interesting

and filled with pleasure in all seasons.

Contents

Acknowledgments

Though neither of them may know it, Edward Miller and Bob Gress were the ones who started me down the path that eventually led to this book. If Ed had not asked me to do *A Pocket Guide to Kansas Flint Hills Wildflowers and Grasses,* and if Bob had not been such an encouraging and enjoyable editor to work with, I would never have dreamed of publishing this book on grasses. I want to thank them both for that pleasurable experience.

Another powerful stimulus for this book has been the days spent afield with good friends. Louise Summers never lacks enthusiasm for a wildflower adventure, whether it means leaving at 5 AM or enduring the blazing heat of a summer afternoon and stiff prairie breezes of 35 MPH; also, Jeff Hansen whose idea of fun is to stand ankle-deep in muck looking for elusive wild plants; Carol Peterson is always ready at a moment's notice to head out and explore new back roads and out-of-the-way prairies; and my dear friend, Marge Streckfus, who has a perpetual zest for life, learning, and the great outdoors. I especially want to thank these people for their indulgence, inspiration, and all those times and discoveries.

It was one fall day while hiking a trail at Tallgrass Prairie National Preserve when another friend, Susan Reimer, and I stopped to admire and take pictures of the tiny flowers of big bluestem that something in my head clicked, telling me that a phenomenon this special needs to be shared. That was the beginning of this book. Through this book, I hope others may enjoy the extraordinary experience of learning about grasses.

In the long journey from concept to printed reality, many people generously lent their time and expertise. From the first burst of interest through the drudgery of revisions, Nancy Goulden was a mainstay, advocate, and quiet counselor. Nancy generously reviewed the manuscript at various stages. Justin Thomas also examined the descriptions. Their interest and honesty is much cherished and added greatly to the final draft. To them my heartfelt thanks.

The most daunting part of this book was designing a simple and useful finding list. Several special friends acted as testers and mentors in working toward a list that would be functional to users. For this I thank Anne Cully, Michele Funston, Brad Guhr, Paula Matile, Sherry Osland, and especially Gene Towne.

I would like to thank my editor, Fred Woodward, for being an essential guiding light throughout the process. His faith in the project from the beginning to end is much appreciated.

Great respect goes to the Interlibrary Loan Department of the Abilene Public Library for obtaining numerous references for me. Big hugs go to Alice Betz for the photograph that follows.

And not least, my gratitude for love and support from my husband, Ken, who always provides helpful suggestions whenever asked, traveled with me without question to distant prairies, and tolerated my two year obsession of photographing grasses.

With a little help from my friends. Cousin, Jim Betz, and husband, Ken Barnard, are holding the black cloth for me. Alice Betz, Jim's wife, took this photo of the three of us.

Legacy of the Heartland

Our genesis was in grassland; perhaps our Garden of Eden was prairie.

JOHN MADSON, *Where the Sky Began*, 2004

Next in importance to the divine profusion of water, light, and air, those three great physical facts which render existence possible, may be reckoned the universal beneficence of grass. . . . Should its harvest fail for a single year, famine would depopulate the earth.

JOHN JAMES INGALLS (Kansas Senator, 1873–1891), "In Praise of Blue Grass," printed in the old *Kansas Magazine*, 1872

The vegetation of Oklahoma, Kansas, and Nebraska comprises more than 3,000 species of wild plants. Of that huge variety of plants, only two out of every 10 species are grasses. Of all the grasses, a mere handful comprises the bulk of the vegetation. In some prairies just one or two grasses can account for 80 to 90 percent of the ground cover, though there may be 200 plant species within the same area.

Those numbers are rather staggering. It is those major wild grasses (big bluestem, little bluestem, Indiangrass, switchgrass, buffalograss, sideoats grama, and blue grama) plus others that may not be found in large numbers but are widespread and easily noticed that are covered in this guide. This book is about wild grasses of Oklahoma, Kansas, and Nebraska, either native or naturalized.

Knowledge of individual grasses is of interest to many people for various reasons. Native grasses have become popular in city, commercial, and home landscaping. There is growing attention to prairie restoration by private individuals, agencies, and organizations. Ranchers and home owners want to learn about prominent or weedy grasses. Naturalists and leaders of youth groups can also find useful information in this book.

The grass plant is truly an amazing wonder often overlooked or taken for granted. Frequently people are surprised to learn that grasses have flowers, just as sunflowers and wild roses; the only difference is that grass flowers are small, inconspicuous, and differ slightly in structure.

Through thoughtful examination the study of grasses is rewarding. Learning about grasses expands the connectedness humans experience with the natural world around us. I hope that I can impart a joy and enthusiasm for these unique plants to readers of this guide.

The Prairies of Oklahoma, Kansas, and Nebraska

Two hundred years ago the central part of the North American continent, from southern Canada into Texas, was covered by wild grasses. From the foothills of the Rocky Mountains toward the east, shortgrass prairies prevailed with blue grama and buffalograss the primary grass cover. Here, with 10 to 22 inches of precipitation, the climate is semiarid. Growing 4 to 24 inches tall, the plants are adapted to infrequent precipitation, high winds, high temperatures, and low humidity.

To the east of these shortgrass prairies, mixedgrass prairies are found where the average annual precipitation is 22 to 30 inches. The characteristic grasses are little bluestem, sideoats grama, and western wheatgrass, which grow two to five feet tall.

Farther east, the tallgrass prairies receive more than 30 inches of annual precipitation, and big bluestem is the dominant grass species, towering to as much as eight feet tall. These three prairie boundaries are not distinct or stable. During drought, the vegetation patterns shift east, but with plentiful moisture the prairie vegetation types reverse, moving west. These three grassland types are divided again into many physiographic regions.

In Nebraska the panhandle is where shortgrasses are found, with tallgrasses in the eastern third. The central part of the state has mixedgrass and a large expanse of sandhills. The Nebraska Sandhills is a unique physiographic region with a distinctive vegetation of its own. Covering over ¼ of the north central part of the state, the Nebraska Sandhills comprise the largest expanse of grass-covered dunes in the Western Hemisphere.

Kansas is divided almost equally into thirds by the three prairie types. Found in the eastern third, the best-known physiographic region in the state, the Flint Hills, is blanketed by tallgrass prairie. Here steep, rocky, flat-topped hills are covered with lush prairie grasses and wildflowers. Although tallgrass prairie once stretched continuously from Canada to Texas, covering an estimated 400,000 square miles (an area nearly five times the entire state of Kansas), the Flint Hills is now the largest remaining tract of native tallgrass prairie in North America.

The Oklahoma tallgrass prairie reaches the state's eastern border only in small pockets or glades. A slice, one or two counties wide, bordering eastern Oklahoma is predominantly oak-hickory forest. From those forested counties, the tallgrass prairie ranged west toward the center of the state. The largest ecosystem type in

Shortgrass prairie in May. Photo taken at Smoky Valley Ranch in Logan County, Kansas, by Ken Barnard.

Late September in the Smoky Hills mixedgrass prairie near Brookville, Kansas. Canada wildrye is in the foreground.

Nebraska Sandhills in August near Purdum. Sandreed is in the foreground.

A Nebraska Sandhills blowout.

Kansas Flint Hills at Tallgrass Prairie National Preserve in May.

Oklahoma was the Cross Timbers, a unique habitat much remarked on by early travelers, comprising short oaks and thickets of brambles and vines intermixed with grasses. The Cross Timbers extended north to south, mingling in and through the tall and midgrasses. To the west of the Cross Timbers, a band of open, mixedgrass prairie once dominated the landscape before reaching the shortgrasses of the Oklahoma panhandle.

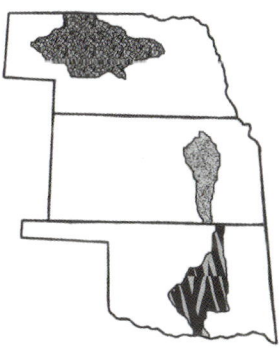

Nebraska Sand Hills

Kansas Flint Hills

Oklahoma Cross Timbers

Oklahoma Cross Timbers in September. Prairie grasses at Keystone Ancient Forest.

Prairie glades and savannas are common in the Oklahoma Cross Timbers. Big bluestem is in the foreground.

Humans have altered the original landscape of these three states, changing the prairie from native perennials to cultivated annuals. Grasses still dominate the landscape, but now they are cultivated fields of wheat, corn, and sorghum.

Due to the radical and rapid changes that North American prairies have undergone over the past century and the resulting loss or degradation of native prairie, many individuals and organizations are interested in prairie restoration. Prairie restoration is the reintroduction of native plant communities intended to reconstruct an ecosystem similar to that which originally existed. When planning for successful prairie restoration, knowledge of plant species and local soils in the area is vital. The process of total restoration is extremely complex and infinitely slow, but early results are rewarding. The experience of watching the plant communities develop each year and seeing the native grasses return and flourish offers inspiration.

Volunteer wheat along the roadside.

The aesthetic appeal of grasses is unique. The varied textures and forms of the grasses adapt them to many uses. If picked early so they won't shatter, many of the grasses make beautiful dried arrangements. In landscaping the grasses are attractive and interesting in all seasons. Stately grasses catch the eye when used in gardens, adapting to a wide range of conditions and requiring minimal maintenance. Grasses have extraordinary diversity, durability, and versatility. Selecting plants that are native to an area ensures that they will be better adapted and take less care, water, and energy. Native grasses can be used in formal garden plantings or in more natural landscaping.

It was pleasant to compare the first tender signs of
the infant year just peeping forth with the stately
beauty of the withered vegetation which had
withstood the winter,—life-everlasting, goldenrods,
pinweeds, and graceful wild grasses, more obvious
and interesting frequently than in summer even.

HENRY DAVID THOREAU, *Walden*, 1854

The prairie turns a kaleidoscope of color.

Little bluestem

Switchgrass

Above: Fall landscape; *below:* winter ice

Indiangrass in January

Switchgrass and rough dropseed

Why Are Grasses Important?

We humans have an intimate daily relationship with grasses. You probably ate grass for breakfast today. Grasses are staple foods eaten worldwide—rice, wheat, oats, and corn are all cultivated grasses. Add cane sugar, another grass, and Americans consume a lot of grass. Beer is made from barley. Other alcoholic beverages are made from rye, wheat, corn, and rice. In fact, by weight 70 percent of all crops are grasses.

Cultivation of grasses as food began about 10,000 years ago, but grasses have many other uses. Some grasses have been highly valued for thousands of years for the construction of homes and furniture, and also in the production of fiber and paper. More recently grasses have been used in biofuel production and even decorative flooring. Turf grasses, ornamental, and lawn grasses are central features of the modern urban American landscape.

Estimates vary as to when grasses first appear in the fossil record. It was sometime near the end of the Cretaceous and beginning of the Tertiary period, 65 to 55 million years ago. This is when flowering plants gained importance. Today the grass family (*Poaceae*) dominates all other flowering plant families in number of individual plants and in widespread distribution. Found on all continents, even in the Antarctic, and in practically all habitats, grasses come in fourth in number of species with 10,000, exceeded only by the aster family with nearly 23,000 species, the orchid family with 22,000, and the bean family with over 19,000 species, according to the Missouri Botanical Garden (listed in the bibliography).

It is generally estimated that grasslands comprise 20 to 30 percent of the land vegetation covering the earth. In both the heart of the city and in remote country, in backyards and along roadsides, a variety of wild grasses can easily be found.

This universal availability of grasses has been beneficial to people. Wild grasses have been fashioned to serve many purposes, from fishing rods and water pipes to rafts, brooms, and basket making. Reeds of woodwind musical instruments are traditionally made from grass. The principal source is *Arundo donax*, giant cane. Flutes may be made from giant cane or bamboo, an important grass from Asia. Some bamboo species have stems reaching eight inches in diameter and 120 feet in height, looking more like a tree than a grass.

The grains of most wild grasses are edible. Patience is required in gathering them because they are so small. They can be cooked whole or ground into flour. *Echinochloa* and *Setaria* are good examples. Eaten since prehistoric times, *Setaria* is one of the oldest cultivated cereals. Grass rhizomes are also sometimes ground into flour. The rhizome of *Agropyron* has reportedly been used in bread and cakes during times of famine. Some wild grass grains have been used to make fermented beverages.

Grasses are also grown as forage for cattle and other livestock to produce meat for the table. They are the foundation for the important livestock industry. Our wild native prairie grasses are highly valued as rangeland for domestic grazing animals. Composition of plant species plays a vital role in promoting both sustainable native pastureland and effective rangeland management. Sustainable native pastureland and rangeland management are based on changes in plant species composition. The starting point for most range-management decisions is to know the range plants by name and to learn their growth habits, forage value, and response to grazing. Plant species can be categorized as "increasers" or "decreasers" corresponding to their shifts in relative abundance in response to grazing. This plant response is often relative to the range type, but consistent on all ranges in some species.

Wildlife is also dependent on grasses. Immense herds of bison, pronghorn, and elk once inhabited the prairies of Oklahoma, Kansas, and Nebraska, living off the rich, nutritious grasses. All sorts of wildlife today still rely on the wild plants around them to meet their needs. Research shows that as much as 25 percent of the annual forage production in grasslands is consumed by wild grazers. Grasses and other plants provide necessary materials to sustain them.

Many bird species, as well as mice and other small rodents, eat the grass grains or other parts of the grass plant for food. The eastern bluebird, meadowlark, prairie vole, and thirteen-lined ground squirrel, among others, use native prairie grasses for nest building. Native grasses such as little bluestem, big bluestem, and switchgrass provide cover for wildlife throughout most of the year and are especially important during winter.

In number, insects are the largest group of grassland plant eaters. Consuming great quantities of vegetative matter, mostly grasses, grasshoppers are common in prairies—Kansas alone has more than 120 species of grasshoppers. Leaf beetles, thrips, aphids, leafhoppers, flea beetles, stinkbugs, leaf-miners, and spittlebugs are other groups of insects that feed on grasses.

Relationships between insects and native plants can be highly specialized. Some butterflies depend on wild grasses. Most noted are small butterflies called skippers whose larvae feed on certain grasses. Among these are tawny-edged, roadside, and least skipper. The stout-bodied caterpillars make nests from the leaves of the food plant by weaving them together with silken threads. Skipper caterpillars often hide in their nest by day and feed on the host plant by night.

Within its natural ecosystem, each grass contributes to the dynamics of natural succession and the grassland ecology. Even if no more than a native weedy colonizer on barren or disturbed prairie soils, within its natural ecosystem every grass has its place and function.

Unique Features and Structure of Grasses

Grasses are widespread because they have adaptations that help them withstand stresses caused by drought, wind, fire, and grazing. When stressed by heat or drought, many grasses roll the edges of their leaf blades inward to reduce water loss. Flat leaves can become threadlike when they are tightly rolled. Dense hairs on the leaves also protect from water loss. Grasses withstand fire and grazing because the growing point is near the base of the leaf or shoot. When clipped or burned, a grass grows back easily from the base. Other types of plants grow from the shoot tip, making regrowth more difficult after clipping.

Grass stems contain a large amount of silica. Particles of silica in the outer cell wall keep grass stems tough and firm, yet strong and resilient. Grass stems and leaves that you see aboveground are often a small percent of the total living weight of the plant. In many grasses half or more of the plant by weight is belowground in the roots. This belowground stored energy helps grasses survive grazing and fire. Grass roots and rhizomes prevent soil erosion by wind and water. Most of the rhizomes are in the upper four or five inches of soil. The total length of cordgrass rhizomes per square foot of soil surface was found to be 75 to 85 feet. Using these numbers, one acre of cordgrass sod may contain as much as 700 miles of soil-binding rhizomes. The network of roots brings up nutrients from deep in the soil. The fertility of healthy grassland soils constantly increases. As plants die they enrich and build soil.

Grasses are anemophilous, or "wind-loving." Rather than relying on insects or animals to carry their pollen, grasses make use of the wind, shedding large amounts of pollen. The two-celled anthers are attached near their middle to the end of a slender filament. Trembling in the wind, they easily discharge the smooth, spherical pollen grains. A single anther of rye is estimated to contain 20,000 pollen grains. I have seen vehicles speeding past smooth bromegrass in flower along roadsides send huge clouds of yellow pollen swirling high in the air. The airborne pollen of some grasses is a source of hay fever. The worst offenders in the central United States are orchardgrass, bluegrass, tall fescue, western wheatgrass, and Johnsongrass.

With no need to attract insects for pollination, grass flowers lack fragrance, nectar, and petals. The small, nonpetaled flowers are nevertheless colorful. Feathery stigmas can be silvery white to yellow or pinkish lavender to deep blue-purple. Anthers range in brilliant colors from greenish yellow to orange and crimson and from lavender to deep purple.

Grasses have many methods of seed dispersal. Seeds are spread by rodents, birds, and other animals that use them as food. Seeds may be buoyant and can be carried along the surface of running water. Some grasses have seeds that catch in the fur of passing animals to be carried great distances. Other grass seeds are light with hairs attached that help them float in the wind, or sometimes the whole seed head breaks and tumbles in the wind, scattering seeds along the way. A few grass seeds are sticky when wet, allowing them to adhere to passers-by.

A clever mechanism possessed by some grasses actually buries the seed deeply in the soil. The seeds have awns that repeatedly twist as they dry and untwist when moistened, drilling the sharply pointed seed into the soil. This process is called "hygroscopicity." The ability of grass seeds to work their way across and into soil becomes a problem when seeds burrow into an animal's fur and skin, creating sores, or a frustration to people who return from a summer prairie hike with socks full of grass "stickers" that are difficult to pull out.

GRASS STRUCTURE. Grasses might appear to be only stems and leaves, but their characteristics are unique. They also have their own unique terminology. Many of the features identifying grass are small. A hand lens is helpful when studying grasses.

ROOTS, STEMS, AND LEAVES. All grasses have fibrous, branching roots. Roots are very shallow in some species and reach depths of over 12 feet in others. Some grasses have underground horizontal stems called rhizomes. A **rhizome** may be recognized because it is jointed and bears scales, which are reduced leaves. Aboveground horizontal stems that trail on the ground and root to produce new plants are called **stolons**.

Although there are exceptions, grasses are usually thin, erect plants with round, hollow stems. The stems have conspicuous bulging joints called **nodes**, where leaves are attached.

Grass leaves are alternate on the stem. The base of the leaf, called the **sheath**, wraps around the stem to protect the enclosed young shoot. The sheath of most grasses is open and overlapping along one side. The sheath edge or **margin** may be hairy or smooth. The flat leaf **blade** projects outward from the stem. Grass leaves are parallel veined with a strap-shaped blade. Hairs or glands may be located on the edge, **collar**, and upper or lower surface of the leaf. On the upper leaf surface at the junction of the blade and sheath is a short membranous appendage or a fringe of hairs called a **ligule**. The ligule prevents dirt or water from entering the sheath. At the base of the blade there are sometimes two small, curved outgrowths that clasp the stem called **auricles**, one on each side of the collar.

FLOWERS AND SPIKELETS. Each grass flower head is composed of many small flowers. Grass flowers have no petals, so what you see

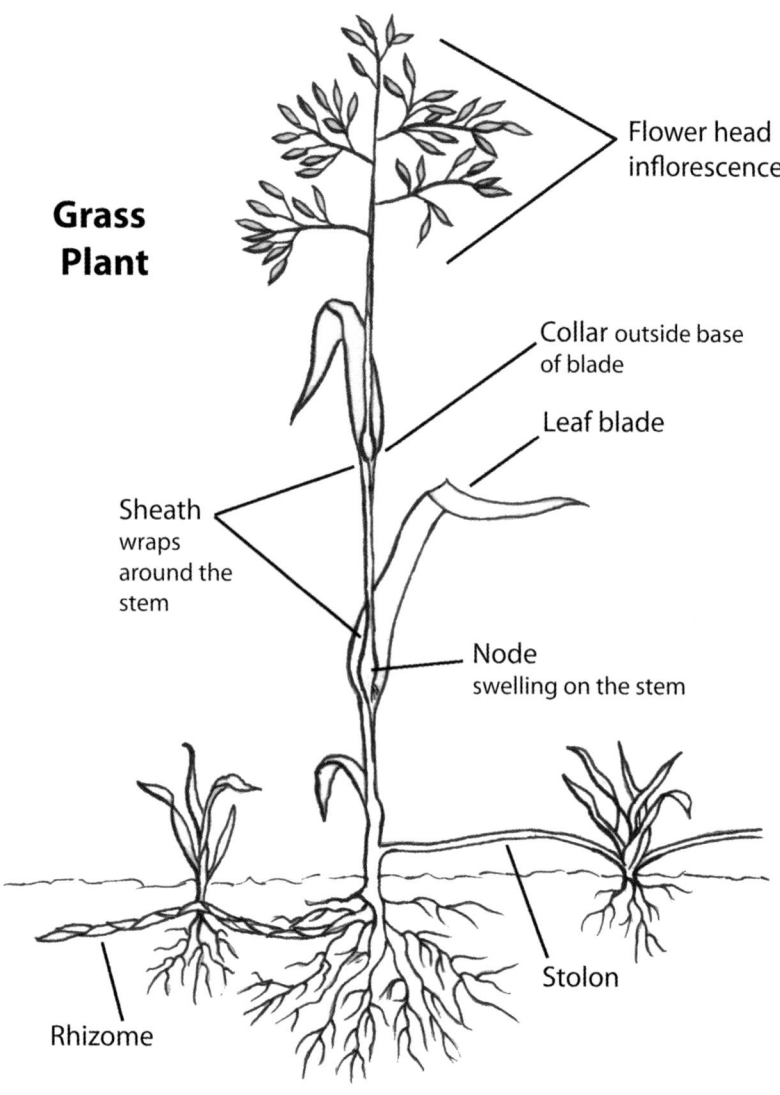

Grass Plant

Flower head inflorescence

Collar outside base of blade

Leaf blade

Sheath wraps around the stem

Node swelling on the stem

Stolon

Rhizome

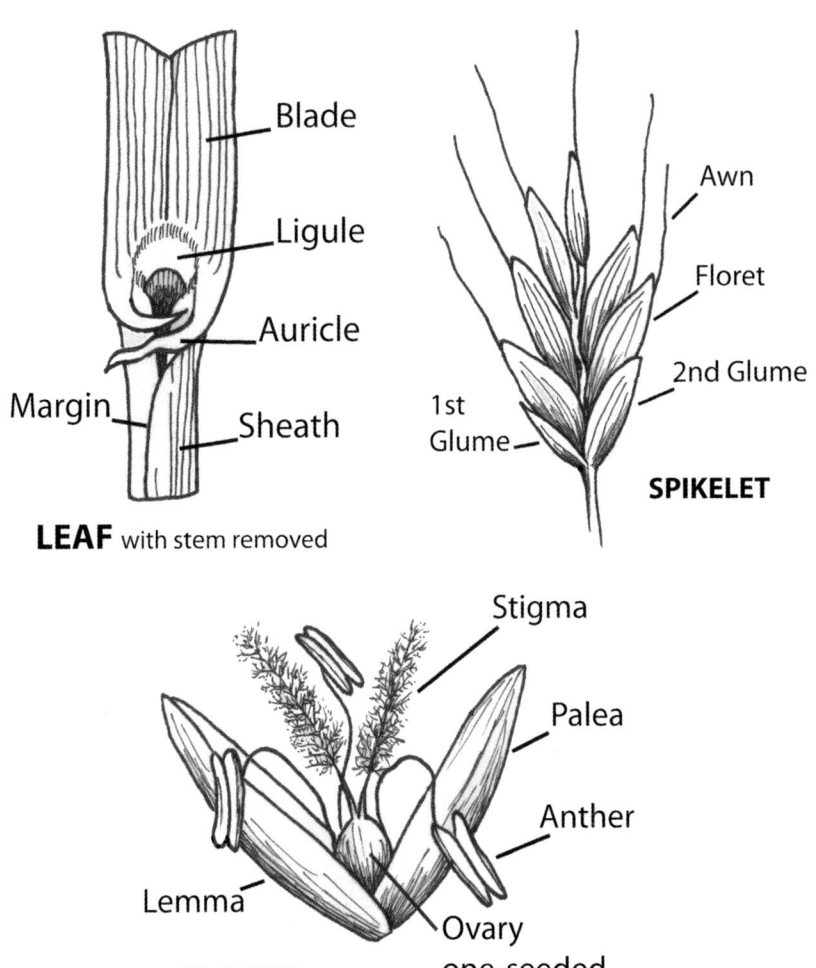

LEAF with stem removed

SPIKELET

FLORET open

are the tiny feathery **stigmas** (normally two) and colorful **anthers** (usually three). These flower parts are enclosed by two protective bracts (**lemma** and **palea**) making up a single **floret**. One to 60 florets are grouped above a pair (infrequently one or both may be reduced or absent) of empty bracts (**glumes**) and collectively called the **spikelet**. Bristles extending from the tip of scales or bracts are called **awns**. Awns vary in length and thickness and may be straight, bent, or twisted. The entire flower head is termed an **inflorescence**.

TYPES OF INFLORESCENCES. The first thing you are likely to notice about most mature grasses is the flower head (inflorescence). The standard botanical terminology for the three basic types of grass heads are spike, raceme, and panicle. There are many different combinations of these three inflorescence types. The outward appearance of each inflorescence type may be very different, since the panicle in one species may be open and spreading and in another closed and narrow.

Some type of organization or simple key is useful in identifying the particular grass in question. Grasses are sometimes divided into subfamilies and tribes to facilitate identification. Keys or finding lists for grass identification can be organized in various ways. Aside from formal taxonomic classification, grasses may be grouped by the type of habitat they occupy, their time of growth, their habits of growth, their height or dominance, their relation to grazing, or any number of other features. In this book I have used the general shape or outline of the inflorescence as an identification aid.

How to Use the Book

The grass owned the prairie. How often I have let my eyes drift like a cloud across long miles of prairie, and not a sight of any shrub from sky to sky! The wild grass had its way. A clump of wild weeds, clouds of wild flowers, but not a bush in sight; for so is the prairie sacred to the grass.

WILLIAM A. QUAYLE, *The Prairie and the Sea,* 1905

In early fall, when folks drive to the mountains to watch the aspens change to gold, I set out over the hills to experience the turning of the grasses. I wade through shoulder-high stands of claret-hued Indian grass, run my fingers through the silvery seed heads of sand bluestem, and nestle into soft clumps of golden switchgrass.

STEPHEN R. JONES, *The Last Prairie,* 2000

The Descriptions

Names

The common name is listed first. A single grass species can have several common names. Common names vary by region and from person to person.

Scientific names are in Latin and are universal around the world. They may honor a person. For instance, the grass *Koeleria* named for the late-eighteenth-century German professor of botany and student of grasses George Ludwig Koeler. However, more often the names describe a particular aspect or characteristic of the plant. For example, *pratensis*, the specific epithet for Kentucky bluegrass, means "of the meadows."

There is only one correct scientific name for each grass, but scientific names are sometimes updated to reflect new knowledge. Both the common names and the scientific names used in this book follow those of PLANTS Database (July 2012).

Names are important because we use them to communicate, but there are many other ways to get to know a grass. So, don't focus only on the name. Look carefully, learn more, and remember what is special to you about each grass.

Height

This may vary somewhat due to seasonal and moisture conditions. Location and soil type also affect the overall height. The heights reported here are from my personal observations.

Native vs. Introduced Grasses

Grasses that have been growing in the prairies and plains of the central United States for a very long time are considered native. They are the species that the first European explorers found when they arrived in the area that has become Oklahoma, Kansas, and Nebraska.

Introduced grasses (nonnatives, sometimes referred to as "naturalized") have increased in number since European settlement. Some of these grasses have been intentionally brought to the area as feed crops or ornamentals, or for erosion control. Their seeds also hitchhike with shipped goods, animals, or travelers. Introduced plants can grow wild without cultivation and are so thoroughly established that they are now part of the flora.

Some introduced species blend into the landscape, but others are invasive. Invasive grasses are undesirable because they are able to compete aggressively with native plants, crowding them out and creating a monoculture. Because invasive plants are a threat to native habitats, agriculture, and public health, they may be designated as noxious by some states. Then they must be controlled or eradicated.

Growth: Life Span and Season of Growth

Grass growth may be annual or perennial. Annual grasses complete their life cycle in one growing season. After producing seeds, the plants die and regenerate from seed in following years. Annuals are opportunistic species. According to weather and other conditions, they may be found in great numbers one year and be absent or few the next. Annuals colonize soil that is bare due to some type of disturbance. They may appear abundantly at a location one year and be replaced by different species the next.

Perennial grasses live three or more years. Perennial plant tops die during the winter, but the underground parts remain alive to produce regrowth the following spring. Most prairie plants are perennials. It is thought that some individual grass plants, such as big bluestem, may be hundreds of years old.

Grasses are classed as either cool-season or warm-season. Cool-season grasses are at their best early in the spring when the temperatures are mild. Most will have completed their life cycles by June. The plants go semidormant in summer and green up again in the fall. In some years, especially in the southern range or with protection, leaves near the ground will stay green all winter. An example is Scribner's panicum and related species of *Dichanthelium*.

Warm-season grasses begin growth in late spring after soil temperatures have warmed to 60 to 65 degrees F. Warm-season grasses dominate in North American prairies and are their showiest in late summer and fall when the plants mature.

Habitat

The kind of place or environment where a plant grows is its habitat. Some grasses require wet or moist soil, while others thrive on dry, rocky sites. Habitat information gives clues about where to look for a certain grass and can be one of the keys to identification.

Description and Comments

The descriptions that accompany each grass species are provided to aid in identification. Descriptions of the plants are not comprehensive but focus only on the characters that distinguish one grass from the others. Some identification features are microscopic and these are avoided. Specialized terminology is kept to a minimum.

To limit technical jargon, words in common usage such as "seed" were chosen. However, grasses have a specialized seed correctly called a grain or caryopsis, referring to the dry, single-seeded fruit with the seed fused to the fruit wall. Other groups of plants also have distinct names for the reproductive part commonly referred to as a seed. In this guide the words "grain" and "seed" are used interchangeably.

Likewise, the word "inflorescence" is preferred by botanists, rather than "flower head," in order to avoid confusion with another plant group. The composite plants, to which sunflowers belong, form dense flower clusters that are termed "heads." These outwardly subtle differences may not be of interest to beginning botanists, but they become important in advanced plant study. If the meaning of a word is not understood, please refer to the glossary.

Other information of interest related to the plant and how it fits into our lives and its environment is also included in the comments.

Geographical Distribution

It is helpful to know the geographic distribution of a particular grass. Most of these common grasses have ranges far beyond the three-state area. Because their expanded range may be of interest or useful to travelers and individuals in other states, continental US maps showing species distribution are provided.

States where the species has been found have lighter shading. The darker shading is the area where the species is most common. Distribution maps are adapted from J. T. Kartesz's North American Plant Atlas (listed in the bibliography).

Photographs

No description can replace actually seeing what a plant looks like. Detailed photographs show important features of each grass. Most of the photos in this book were taken when the plants were blooming. When comparing a grass with the photographs, keep in mind that grasses will change during the seasons and may vary by region or because of soil or moisture conditions. To visually clarify the leaf size, a close-up picture of a wider leaf of the species with a dime ($^{11}/_{16}$ inch) as reference is shown for each grass. *A bracket representing ¼ inch accompanies smaller features.*

Grass Finding Lists

These lists are included to aid the user in the identification of grasses. Plants are easiest to identify when the whole plant is available for examination. This means flowering heads, stems, leaves, and roots. Descriptions in this guide point out characters that are visible without magnification, but a hand lens exposes the intricate beauty and delicate details otherwise missed. Learning to use the hand lens will help you expand your study of grasses.

Grasses in this book are first grouped by the shape or outline of the flower head (the inflorescence). Plants, of course, look different at different stages in growth. Study the grass when it is flowering because as it develops the head changes, in several species spreading just before flowering and contracting later as the seeds form. Others, such as silver beardgrass, are contracted during flowering, and then the head expands as seeds ripen. If you don't find your plant in one group, check another that is similar. And remember that these are the 70 grass species that you will find most often, but many other grasses may be encountered that are not in this guide.

Use the following finding list to decide in which of the seven sections of the book your grass may be found. Knowing the season when your grass is growing and blooming will help to narrow the identity. Plant height is also a clue. Finally the grasses are arranged alphabetically by the scientific name in order to group similar related plants.

Index to the Grass Groups

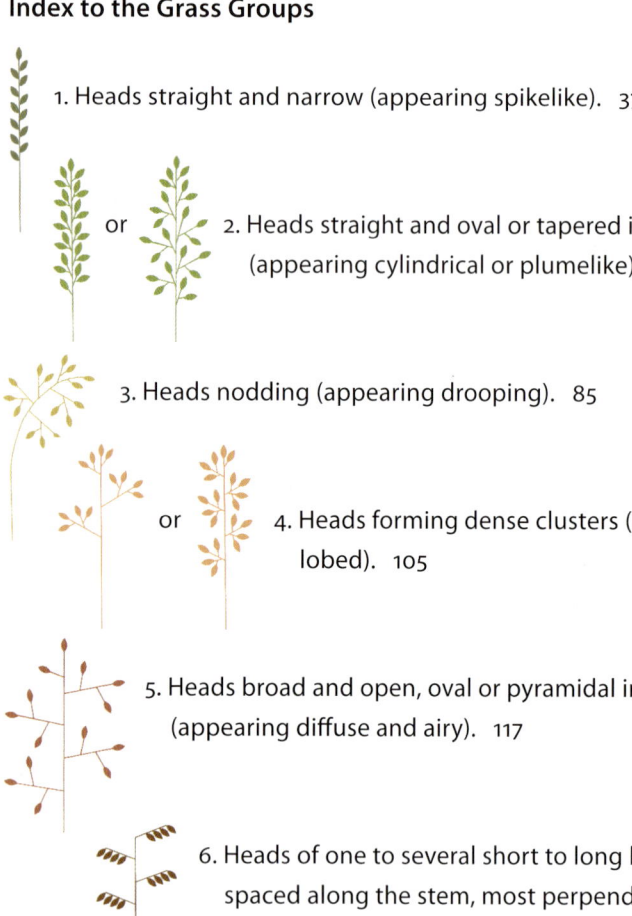

1. Heads straight and narrow (appearing spikelike). 37

or
2. Heads straight and oval or tapered in outline (appearing cylindrical or plumelike). 57

3. Heads nodding (appearing drooping). 85

or
4. Heads forming dense clusters (appearing lobed). 105

5. Heads broad and open, oval or pyramidal in outline (appearing diffuse and airy). 117

6. Heads of one to several short to long branches spaced along the stem, most perpendicular to the stem (appearing comblike or ladderlike). 173

7. Heads of two to many narrow branches radiating in whorls near the top of the stem or along the stem (appearing fingerlike or spokelike, as a wheel). 193

8. Heads low and hidden by the leaves (not immediately apparent). 177

1. Heads straight and narrow (appearing spikelike)
    1A. Cool-season (early spring until June)
        A1. Shortgrass (4–24 inches)
            jointed goatgrass (*Aegilops cylindrica*) 38
            sixweeks fescue (*Vulpia octoflora*) 53
        A2. Midgrass (2–4 feet)
            western wheatgrass (*Pascopyrum smithii*) 44
    1B. Warm-season (summer and early fall)
            prairie cupgrass (*Eriochloa contracta*) 41
            little bluestem (*Schizachyrium scoparium*) 47
            rough dropseed (*Sporobolus compositus*) 50

2. Heads straight and oval or tapered in outline (appearing cylindrical or plumelike)
    2A. Cool-season (early spring until June)
        A1. Shortgrass (4–24 inches)
            little barley (*Hordeum pusillum*) 67
        A2. Midgrass (2–4 feet)
            prairie Junegrass (*Koeleria macrantha*) 70
            prairie wedgegrass (*Sphenopholis obtusata*) 82
            reed canarygrass (*Phalaris arundinacea*) 73
    2B. Warm-season (summer and early fall)
        B1. Shortgrass (4–24 inches)
            prairie threeawn (*Aristida oligantha*) 121
            saltgrass (*Distichlis spicata*) 111
            stinkgrass (*Eragrostis cilianensis*) 64
        B2. Midgrass (2–4 feet)
            Caucasian bluestem (*Bothriochloa bladhii*) 58
            silver beardgrass (*Bothriochloa laguroides*) 61
            barnyardgrass (*Echinochloa muricata*) 114
            little bluestem (*Schizachyrium scoparium*) 47
            yellow foxtail (*Setaria pumila*) 76
        B3. Tallgrass (4–7 feet)
            Indiangrass (*Sorghastrum nutans*) 79

3. Heads nodding (appearing drooping)

   3A. Cool-season (early spring until June)

      A1. Shortgrass (4–24 inches)

         ticklegrass (*Agrostis hyemalis*) 118

         downy brome (*Bromus tectorum*) 86

         foxtail barley (*Hordeum jubatum*) 99

      A2. Midgrass (2–4 feet)

         Canada wildrye (*Elymus canadensis*) 89

         fowl mannagrass (*Glyceria striata*) 92

         porcupinegrass (*Hesperostipa spartea*) 95

   3B. Warm-season (summer and early fall)

      green bristlegrass (*Setaria viridis*) 102

      purpletop tridens (*Tridens flavus*) 170

4. Heads forming dense clusters (appearing lobed)

   4A. Cool-season (early spring until June)

      orchardgrass (*Dactylis glomerata*) 108

   4B. Warm-season (summer and early fall)

      B1. Shortgrass (4–24 inches)

         sandbur (*Cenchrus longispinus*) 106

         saltgrass (*Distichlis spicata*) 111

      B2. Midgrass (2–4 feet)

         barnyardgrass (*Echinochloa muricata*) 114

5. Heads broad and open, oval or pyramidal in outline (appearing diffuse and airy)

   5A. Cool-season (early spring until June)

      A1. Shortgrass (4–24 inches)

         ticklegrass (*Agrostis hyemalis*) 118

         Scribner's panicum (*Dichanthelium oligosanthes*) 130

      A2. Midgrass (2–4 feet)

         smooth brome (*Bromus inermis*) 124

         fowl mannagrass (*Glyceria striata*) 92

rice cutgrass (*Leersia oryzoides*) 144

Kentucky bluegrass (*Poa pratensis*) 159

tall fescue (*Schedonorus arundinaceus*) 162

5B. Warm-season (summer and early fall)

B1. Shortgrass (4–24 inches)

prairie threeawn (*Aristida oligantha*) 121

fall witchgrass (*Digitaria cognata*) 133

stinkgrass (*Eragrostis cilianensis*) 64

Carolina lovegrass (*Eragrostis pectinacea*) 136

scratchgrass (*Muhlenbergia asperifolia*) 147

witchgrass panicum (*Panicum capillare*) 150

B2. Midgrass (2–4 feet)

Caucasian bluestem (*Bothriochloa bladhii*) 58

barnyardgrass (*Echinochloa muricata*) 114

stinkgrass (*Eragrostis cilianensis*) 64

purple lovegrass (*Eragrostis spectabilis*) 138

sand lovegrass (*Eragrostis trichodes*) 141

fall panicgrass (*Panicum dichotomiflorum*) 153

sand dropseed (*Sporobolus cryptandrus*) 168

purpletop tridens (*Tridens flavus*) 170

B3. Tallgrass (4–7 feet)

prairie sandreed (*Calamovilfa longifolia*) 127

switchgrass (*Panicum virgatum*) 156

Johnsongrass (*Sorghum halepense*) 165

6. Heads of one to several short to long branches spaced along the stem, most perpendicular to the stem (appearing comblike or ladderlike)

6A. Shortgrass (4–24 inches)

buffalograss (*Bouteloua dactyloides*), male 177

blue grama (*Bouteloua gracilis*) 180

hairy grama (*Bouteloua hirsuta*) 183

tumblegrass (*Schedonnardus paniculatus*) 188

6B. Midgrass (2–4 feet)

      sideoats grama (*Bouteloua curtipendula*) 174

      sand paspalum (*Paspalum setaceum*) 185

6C. Tallgrass (4–7 feet)

      prairie cordgrass (*Spartina pectinata*) 190

7. Heads of two to many narrow branches radiating in whorls near the top of the stem or along the stem (appearing fingerlike or spokelike, as a wheel)

7A. Shortgrass (4–24 inches)

      windmillgrass (*Chloris verticillata*) 197

      hairy crabgrass (*Digitaria sanguinalis*) 200

      goosegrass (*Eleusine indica*) 203

7B. Tallgrass (4–7 feet)

      big bluestem (*Andropogon gerardii*) 194

      eastern gammagrass (*Tripsacum dactyloides*) 206

8. Heads low and hidden by the leaves (not immediately apparent)

      buffalograss (*Bouteloua dactyloides*), female 177

The Grasses

To the casual observer the grasses are but "grass," and to few is their diversity, their beauty, and their value apparent.

> MARY FRANCIS BAKER, *The Book of Grasses,* 1912

Nature is an open book for those who care to read. Each grass-covered hillside is a page on which is written the history of the past, conditions of the present, and predictions of the future. Some see without understanding; but let us look closely.

> JOHN WEAVER, *North American Prairie,* 1954

Group 1

Heads straight and narrow (appearing spikelike)

jointed goatgrass (*Aegilops cylindrica*)
prairie cupgrass (*Eriochloa contracta*)
western wheatgrass (*Pascopyrum smithii*)
little bluestem (*Schizachyrium scoparium*)
rough dropseed (*Sporobolus compositus*)
sixweeks fescue (*Vulpia octoflora*)

Jointed Goatgrass

Jointed Goatgrass

Aegilops cylindrica

HEIGHT: 10–32 inches

GROWTH: Winter annual introduced from Eurasia, cool-season

HABITAT: Fields, waste places, and roadsides

DESCRIPTION: Stems of goatgrass are branched from the base. The sheath is hairy on the margin. Narrow blades have clasping auricles at the base. Flower heads are distinctively cylindrical with spikelets fitting into notches on the stem. Spikelet awns are often over 1 inch long.

COMMENTS: The hard spikelets are the source of the genus name, *Aegilops*, from "aegis," the armor plate once used in battle to protect the chest. Goatgrass is very similar to wheat and will hybridize with it. This grass is a weed in grain fields.

Inflorescence

Collar and node

Blade

Jointed Goatgrass

Above: Reddish plants in the background are goatgrass in late May

Left: Seeds break apart and germinate in fall

Prairie Cupgrass

Eriochloa contracta

HEIGHT: 12–32 inches

GROWTH: Native annual, warm-season

HABITAT: Open, moist soils; often on disturbed roadsides

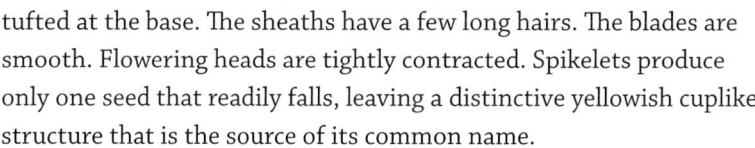

DESCRIPTION: Prairie cupgrass is densely tufted at the base. The sheaths have a few long hairs. The blades are smooth. Flowering heads are tightly contracted. Spikelets produce only one seed that readily falls, leaving a distinctive yellowish cuplike structure that is the source of its common name.

COMMENTS: This species is native to Texas, Oklahoma, Kansas, and Nebraska. It is introduced elsewhere in the United States. Worldwide there are about 30 species of *Eriochloa*, mostly in tropical and subtropical regions.

Prairie Cupgrass

Prairie Cupgrass

Leaf blade

Inflorescence

The "cups" are visible
after seeds have fallen

Western Wheatgrass

Western Wheatgrass

Pascopyrum smithii

HEIGHT: 1–3 feet

GROWTH: Native perennial, cool-season

HABITAT: Roadsides and pastures, it thrives in heavy clay or claypan soils

DESCRIPTION: Western wheatgrass forms large colonies with creeping underground stems (rhizomes); these rhizomes prevent soil erosion. Plants have a waxy, bluish-green color. The sheaths are round and smooth. Each blade is strongly ridged above with two clasping projections (auricles) where it joins the stem. Blade edges tend to roll inward when dry. The overlapping spikelets are ⅜ to ¾ inch long.

COMMENTS: Western wheatgrass furnishes high protein hay for livestock. The seeds, produced in August, are food for songbirds, upland game birds, and many small mammals.

Plants have a bluish color

Western Wheatgrass

Auricles and open sheath

Flowering

Blade with ridges

Rhizome

Little Bluestem

Schizachyrium scoparium

HEIGHT: 2–4 feet

GROWTH: Native perennial, warm-season

HABITAT: Upland prairies and roadsides

DESCRIPTION: Roots may reach depths of 8 feet. Stems are slightly flattened, waxy,

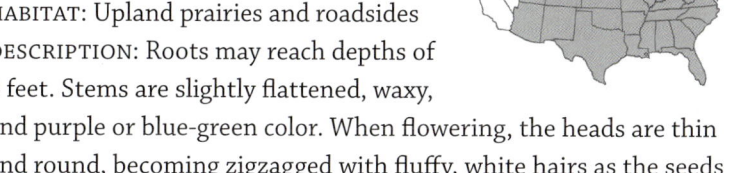

and purple or blue-green color. When flowering, the heads are thin and round, becoming zigzagged with fluffy, white hairs as the seeds mature. Awns up to ½ inch long are spirally twisted and bent near the bases.

COMMENTS: The plants become pinkish red after frost and remain standing all winter, providing good shelter for wildlife. Little bluestem is larval host to common wood nymph, Leonard's skipper, ottoe skipper, and crossline skipper. Little bluestem is an important component in restoration seed mixes. It is the state grass of Kansas and Nebraska.

Bluish-purple color of the stems

Maturing seeds on the left and flowering branch on the right

Leaf blade

Little Bluestem

Little Bluestem

Far left: Plant with flowers

Left: Plant producing seeds

Broomsedge bluestem

SIMILAR SPECIES: Broomsedge (*Andropogon virginicus*) is a conspicuous grass in the southern part of the three-state region and resembles little bluestem in stature. The upper stems branch, making this species appear leafier, and the fall color is more orangish. Broomsedge grows throughout Oklahoma (except in the panhandle) and in the eastern ⅔ of Kansas, and is absent in Nebraska.

Rough Dropseed

Rough Dropseed

Sporobolus compositus

HEIGHT: 2–4 feet

GROWTH: Native perennial, warm-season

HABITAT: Dry soil of upland prairies, road-sides, and railroads

DESCRIPTION: Rough dropseed is leafy and tufted at the base, with erect, often solitary stems. Lower blades are long and narrow, each tapering to a threadlike point. The slim, crowded flower heads often are partly or completely hidden in the sheaths.

COMMENTS: During the winter plants turn whitish and the broad upper sheath that encases the flowers opens, waving in the wind and giving this plant the name "flaggrass." The species is widely distributed. Forage value is fair. It is reported that grains of the *Sporobolus* species were ground into flour by the Kiowa Indians.

Inflorescence partly enclosed in the sheath

Blade and short ligule

Rough Dropseed

"Flags" in winter

Sixweeks Fescue

Vulpia octoflora

HEIGHT: 5–20 inches

GROWTH: Native annual, cool-season

HABITAT: Dry, upland prairies and sandy, well-drained soil

DESCRIPTION: The narrow and stiffly erect stems of sixweeks fescue do not branch. Leaf blades are usually in-rolled and threadlike. Spikelets are flattened with crowded florets arranged in a herringbone pattern. Each floret has only one stamen. Awns are up to ¼ inch long.

COMMENTS: The name *octoflora*, meaning "eight-flowered," is a misnomer because there are often more than eight florets per spikelet. This grass has low forage value. The seeds are eaten by birds.

Inflorescence

Involute leaf blade

Sixweeks Fescue

Sixweeks Fescue

Plants are stiff and upright

Group 2

Heads straight and oval or tapered in outline
(appearing cylindrical or plumelike)

or

Caucasian bluestem (*Bothriochloa bladhii*)
silver beardgrass (*Bothriochloa laguroides*)
stinkgrass (*Eragrostis cilianensis*)
little barley (*Hordeum pusillum*)
prairie Juneyrass (*Koeleria macrantha*)
reed canarygrass (*Phalaris arundinacea*)
yellow foxtail (*Setaria pumila*)
Indiangrass (*Sorghastrum nutans*)
prairie wedgegrass (*Sphenopholis obtusata*)

Caucasian Bluestem

Caucasian Bluestem

Bothriochloa bladhii
HEIGHT: 2–3 feet
GROWTH: Perennial introduced from
Australia and south Asia, warm-season
HABITAT: Roadsides, pastures, and prairies
DESCRIPTION: Plants grow in clumps. Erect
stems are usually hairy at the nodes. Sheaths and blades lack hairs
except at the blade base. Each blade has a thickened midrib. The central stem of each inflorescence is longer than the branches.
COMMENTS: Caucasian bluestem and yellow bluestem (*B. ischaemum*)
are collectively called Old World bluestems. They have a high tolerance to grazing, high water-use efficiency, and are planted as livestock forage. Unfortunately, attempts to improve native rangeland
with exotic plants often create problems. Intermixed in native
rangelands, both of these grasses are avoided by grazing animals. Old
World bluestems can aggressively outcompete native plants, eventually producing a monoculture that provides little if any wildlife habitat and a less diverse and productive rangeland. Currently, Caucasian
bluestem is established in only three Nebraska counties.

Caucasian bluestem along a roadside

Caucasian Bluestem

Caucasian bluestem inflorescence

Yellow bluestem inflorescence

Caucasian bluestem blade with hairs at the base

Silver Beardgrass

Bothriochloa laguroides

HEIGHT: 16–35 inches

GROWTH: Native perennial, warm-season

HABITAT: Dry soil of upland prairies, roadsides, and waste areas

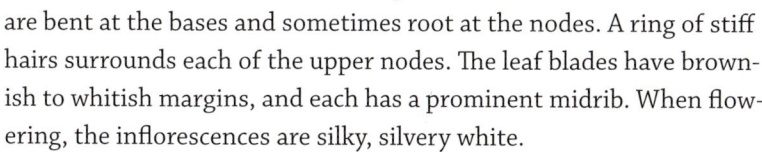

DESCRIPTION: Stems of silver beardgrass
are bent at the bases and sometimes root at the nodes. A ring of stiff
hairs surrounds each of the upper nodes. The leaf blades have brownish to whitish margins, and each has a prominent midrib. When flowering, the inflorescences are silky, silvery white.

COMMENTS: Silver beardgrass increases when range condition is
poor, but decreases as conditions improve. Cottony, white seed heads
of this grass are noticeable along roadsides and in disturbed areas
in late summer and fall. It is found in only a few southern Nebraska
counties.

Silver Beardgrass

Silver Beardgrass

Leaf blade

Head with seeds

Flowering

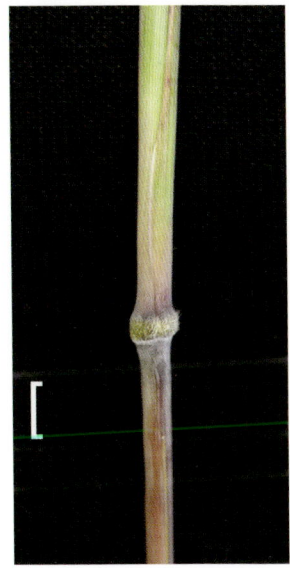

Node and open sheath above

Stinkgrass

Spikelets

Stinkgrass

Eragrostis cilianensis

HEIGHT: 4–20 inches

GROWTH: Annual introduced from Europe, warm-season

HABITAT: Disturbed soils, roadsides, and waste places

DESCRIPTION: Stems of stinkgrass are bent at the bases and then become upright. The sheaths are hairy at the collars. Foliage and spikelets are edged by many small, warty, resinous glands appearing as pitted bumps (a hand lens is helpful). Flatted and awnless spikelets are crowded in the flower heads and have a distinctive grayish-green color.

COMMENTS: Crushed plants have an unpleasant odor. Stinkgrass is avoided by livestock and is thought to be poisonous, especially to horses, if eaten in large quantities.

Warty glands on leaf edges

Stinkgrass

Blade and hairy collar

Little Barley

Hordeum pusillum

HEIGHT: 4–16 inches

GROWTH: Native winter annual, cool-season

HABITAT: Disturbed areas with bare, dry soils and crop fields, especially alkaline soils

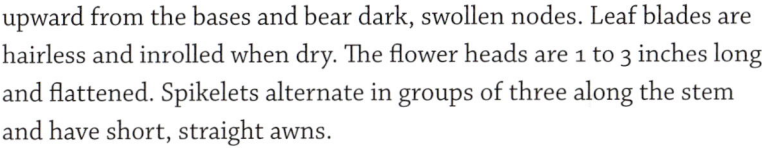

DESCRIPTION: Stems of little barley are bent upward from the bases and bear dark, swollen nodes. Leaf blades are hairless and inrolled when dry. The flower heads are 1 to 3 inches long and flattened. Spikelets alternate in groups of three along the stem and have short, straight awns.

COMMENTS: Little barley, like many annuals, is abundant in some years and at some locations but rare at other times when weather or range management changes. In Europe, straw of the closely related cultivated barley (*H. vulgare*) is used for pulp in papermaking.

Little Barley

Inflorescence

Leaf blade

Flowering

Collar region

Prairie Junegrass

Prairie Junegrass

Koeleria macrantha

HEIGHT: 8–24 inches

GROWTH: Native perennial, cool-season

HABITAT: Upland prairies and dry soils

DESCRIPTION: This tufted plant has stems that are stiffly erect with leaves mostly at the bases. Fine hairs are located on the stems just below the flower heads. The blades are narrow with broad ribs, deep furrows, and prow-shaped tips. When the inflorescences expand to flower, they have a sparkling, silvery-green appearance like an upside-down icicle. Later, the entire plant becomes a golden yellow.

COMMENTS: Junegrass is a common component of high-quality upland prairie and makes excellent forage, but the scattered plants provide low forage production. Prairie Junegrass supplies food for birds and small mammals. This species has many uses by the Cheyenne in their ceremonial customs.

Flowering

Short hairs on the stem below the inflorescence

Prairie Junegrass

Blade with ribs

Open sheath exposing the node

Reed Canarygrass

Phalaris arundinacea
HEIGHT: 2–6 feet
GROWTH: Native perennial, cool-season
HABITAT: Wet soil, marshes, ditches, and
riverbanks
DESCRIPTION: Erect, coarse plants arise
from large, long, scaly rhizomes. The sheaths are smooth and conspicuously air-chambered. The blades are flat, each with a prominent
midrib on the underside. The flower heads are open when flowering,
becoming dense, narrow, and straw colored at maturity. Spikelets are
compressed.

COMMENTS: This highly variable species forms large colonies. It is
typically considered native in North America, but due to the intro-
duction of aggressive European cultivars and subsequent inter-
breeding, native populations may no longer exist. Introduced and
native plants of reed canarygrass are difficult to distinguish from
each other. The invasive nature of nonnative canarygrass is a serious
problem.

Flowering

Ligule

Reed Canarygrass

Reed Canarygrass

Blade and ligule

Rhizomes

Yellow Foxtail

Yellow Foxtail

Setaria pumila

HEIGHT: 8–48 inches

GROWTH: Annual introduced from Europe, warm-season

HABITAT: Cultivated ground, waste places, roadsides, and other highly disturbed sites

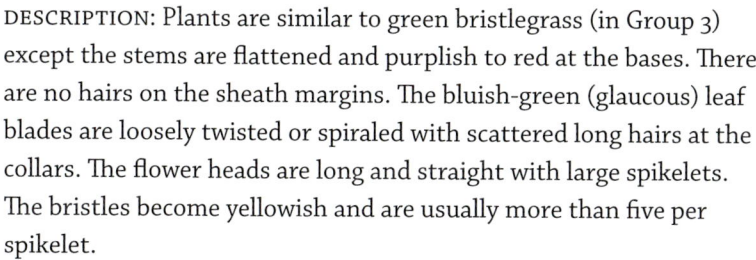

DESCRIPTION: Plants are similar to green bristlegrass (in Group 3) except the stems are flattened and purplish to red at the bases. There are no hairs on the sheath margins. The bluish-green (glaucous) leaf blades are loosely twisted or spiraled with scattered long hairs at the collars. The flower heads are long and straight with large spikelets. The bristles become yellowish and are usually more than five per spikelet.

COMMENTS: The seeds of this grass are one of the most important foods for many species of wildlife. Planted long ago in China and found in great quantities at ancient dwellings, *Setaria* (a type of millet) is among the earliest of cultivated grains.

Inflorescence

Sheath is shorter than the internode

Yellow Foxtail

Stem base

Leaf blade

Indiangrass

Sorghastrum nutans
HEIGHT: 3–7 feet
GROWTH: Native perennial, warm-season
HABITAT: Native prairie uplands in a variety
of soil types
DESCRIPTION: Indiangrass forms clumps
from short rhizomes. Nodes are hairy. The sheaths are open with two
stiff, erect projections adjoining each ligule. Blades are flat and hairless, narrowing at the bases. Dense, yellowish flower heads may be
12 inches long. Florets have hairlike awns that are spirally twisted
and bent near the bases.
COMMENTS: This is an important tallgrass prairie plant that provides
excellent summer forage. Indiangrass is desirable and functional in
home or business landscaping. The showy plumes are attractive in
dried arrangements. Indiangrass is the state grass of Oklahoma.

Flowers

Stem removed showing
"rabbit ears" at top of
sheath and narrow blade
base

Indiangrass

Indiangrass

Collar region with "ears"

Bluish color leaves

Prairie Wedgegrass

Prairie Wedgegrass

Sphenopholis obtusata
HEIGHT: 10–38 inches
GROWTH: Native perennial, cool-season
HABITAT: Prairies in moist soil near ponds, streams, and springs
DESCRIPTION: Plants form small clumps from fibrous roots. Smooth stems are erect with swollen, purplish nodes. Leaf sheaths and blades are often hairy. Flower heads are usually narrow, dense, and light green, turning tan with age. Spikelets are two to three flowered. The first glume is narrow; the second is broadly ovate.
COMMENTS: A good forage grass, but plants are rarely abundant at one location. Prairie wedgegrass might be confused with prairie Junegrass, but the habitat for wedgegrass is usually much wetter than the one for Junegrass. There are four species of *Sphenopholis* in the United States.

Leaf blade

Prairie Wedgegrass

Inflorescence

Ligule and open sheath

Group 3

Heads nodding (appearing drooping)

downy brome (*Bromus tectorum*)
Canada wildrye (*Elymus canadensis*)
fowl mannagrass (*Glyceria striata*)
porcupinegrass (*Hesperostipa spartea*)
foxtail barley (*Hordeum jubatum*)
green bristlegrass (*Setaria viridis*)

Downy Brome

Plants are softly hairy

Downy Brome

Bromus tectorum

HEIGHT: 8–24 inches

GROWTH: Annual or winter annual introduced from Eurasia, cool-season

HABITAT: Roadsides, grain fields, waste
places

DESCRIPTION: A weedy species, plants often occur in dense stands.
Sheaths are softly hairy with short, downward pointing hairs. Flat
blades are covered in soft hairs. Flower heads are drooping and
become purplish as they mature. Awns are ¼ to ¾ inches long.

COMMENTS: Plants begin growth as early as March and produce a lux-
uriant appearance; however, they have a short growing period during
which the forage value is poor to fair, giving it the name "cheatgrass."
Once the plants mature, the long awns can be injurious to grazing
animals. In the fall, soon after this plant germinates, it is heavily
eaten by Canada geese.

Characteristic color in late May

Downy Brome

Blade and short ligule

Inflorescence

SIMILAR SPECIES: Japanese brome (*B. japonicus*) and rye brome (*B. secalinus*) are close in appearance to downy brome. Downy brome comes into flower earlier than any of the other annual bromes.

Japanese brome

Canada Wildrye

Elymus canadensis

HEIGHT: 2–4 feet

GROWTH: Native perennial, cool-season

HABITAT: Mesic to dry upland prairies and roadsides

DESCRIPTION: Stems of Canada wildrye are tufted and unbranched. Stems and sheaths are smooth and often glaucous. Wide blades are flat and hairless with prominent auricles at the bases. The flower heads are arched or nodding with awns ½ to 1 ½ inches long. The awns are curved at maturity.

COMMENTS: Ergot can infest the seed heads, making spikelets irregular shaped and black. Canada wildrye is a desirable grass as forage or hay for livestock, but is rarely abundant in the region. It proves excellent nesting and winter cover for upland game birds and small mammals.

Auricles with stem removed

Flowering

Canada Wildrye

Canada Wildrye

Leaf blade

Curved awns

SIMILAR SPECIES: Virginia wildrye (*Elymus virginicus*) is a highly variable species, similar to Canada wildrye in many respects. It is distinguished by erect heads and spikelets with straight awns. Usually, the awns are shorter than those of Canada wildrye. Virginia wildrye is found in open habitats, but also is frequent in shady wooded locations throughout the region.

Virginia wildrye

Green leaves of *E. virginicus* during winter

Fowl Mannagrass

Fowl Mannagrass

Glyceria striata

HEIGHT: 15–40 inches

GROWTH: Native perennial, cool-season

HABITAT: Wet soil along streams, ponds, and springs

DESCRIPTION: Fowl mannagrass stems are erect. Sheaths are closed to near the top. The blades are flat or folded and firm. Pyramidal-shaped flower heads are open and arched with drooping branch tips. Ovate spikelets, with four to six florets, are awnless. Each floret has only two stamens.

COMMENTS: Seeds are eaten by waterfowl. Plants are of good forage value. *Glyceria* is from the Greek *glykeros*, meaning "sweet," and refers to the sweet taste of the seeds from some species.

Spikelets

Collar and ligule

Fowl Mannagrass

Leaf blade

Blade tips

Porcupinegrass

Hesperostipa spartea

HEIGHT: 2–4 feet

GROWTH: Native perennial, cool-season

HABITAT: Upland prairies

DESCRIPTION: Erect, slender stems of por-
cupinegrass are topped with waving heads.
The sheaths are longitudinally ridged with short hairs on the margins. The blades are strongly ridged above with conspicuous ⅜-inch-long membranes (ligules) at the base of each leaf blade. The spikelet awns are 4 to 6 inches long. When the seeds drop, the 1½-inch-long, white, papery bracts (glumes) are visible at the tops of the plants from quite a distance.

COMMENTS: Porcupinegrass is an important prairie grass that decreases with heavy grazing. The Pawnee, Ponca, and Omaha tribes used the seeds as a hairbrush by bundling the long awns and burning the points off the grains.

Showy glumes when seeds have fallen

Porcupinegrass

Porcupinegrass

Collar region

Blade and ligule

Seeds and awns

Flowers

Porcupinegrass

SIMILAR SPECIES: Needle-and-thread (*Hesperostipa comata*) is a western relative of porcupinegrass, occurring in the western ⅔ of Nebraska, the western ¼ of Kansas, and the Oklahoma panhandle. It is distinguished by shorter glumes, less than 1 ¼ inches long. The long awn is thin and threadlike. It is frequently found in areas with sandy soil and less than 25 inches of annual precipitation.

Needle-and-thread

Threadlike awns and needle-sharp seeds

Foxtail Barley

Hordeum jubatum

HEIGHT: 1–2 feet

GROWTH: Native perennial, cool-season

HABITAT: Temporarily wet areas, such as ditches and low grassy drainages

DESCRIPTION: These short-lived, shallow-rooted plants have slender stems with dark joints. The sheaths are round and hairless. Nodding flower heads are silky, glistening, and sometimes tinted rose purple. The awns are 1 to 2 ½ inches long.

COMMENTS: Awns can injure the mouth, throat, nose, or eyes of grazing animals, causing infections and sores. Awns can also become embedded in the fleece of sheep, reducing the value of the wool. Foxtail barley is an attractive plant often found along roadsides due to its high tolerance for salt and disturbance.

Foxtail Barley

Leaf blade

Foxtail Barley

Inflorescence with long awns

Node

Green Bristlegrass

Green Bristlegrass

Setaria viridis

HEIGHT: 6–36 inches

GROWTH: Annual introduced from Europe, warm-season

HABITAT: Highly disturbed sites, such as roadsides and crop fallow

DESCRIPTION: Green bristlegrass, like most annual grasses, has a shallow root system. Stems are round and tufted at the base. The sheaths are hairy on the margins. Flower heads are upright as flowering begins and soon nodding as they mature. The ovoid florets are rounded and nearly flat on one side. One to three bristles project from the base of each spikelet. This *Setaria* is the earliest to flower and set seeds.

COMMENTS: Seeds of this weedy species are preferred by morning dove, bobwhite, dickcissel, redwing blackbird, grasshopper sparrow, and tree sparrow.

Top left: Blade

Above: Inflorescence

Bottom left: Collar region

Green Bristlegrass

Group 4

Heads forming dense clusters (appearing lobed)

sandbur (*Cenchrus longispinus*)
orchardgrass (*Dactylis glomerata*)
saltgrass (*DIstichlis spicata*)
barnyardgrass (*Echinochloa muricata*)

Sandbur

Inflorescence

Sandbur

Cenchrus longispinus
HEIGHT: 8–32 inches
GROWTH: Native annual, warm-season
HABITAT: Roadsides and waste areas, often
sandy soils
DESCRIPTION: Sandbur is branching and
often sprawling, forming large mats. The stems and sheaths are flattened, frequently reddish-tinged. The blades are sparsely hairy. The
inflorescences are compact clusters of spiny, yellowish burs, each en-
closing two to three spikelets.
COMMENTS: The burs are painful when skin is punctured, but the foliage makes good forage before the burs form. Seeds are spread when
burs cling to animal fur, machinery, or human clothing.

Leaf blade

Flattened sheath

Orchardgrass

Orchardgrass

Dactylis glomerata
HEIGHT: 8–45 inches
GROWTH: Perennial introduced from
Europe, cool-season
HABITAT: Fields, pastures, lawns, and waste
areas

DESCRIPTION: Plants are tufted. The sheaths are compressed and closed to the midpoints. The papery ligules are ¼ to ½ inch long. The soft blades are folded when immature, but are later flat or V-shaped, each with a prominent midvein on the underside. Spikelets form irregular, one-sided clusters on spreading branches.

COMMENTS: In some areas, orchardgrass is cultivated as a hay and pasture grass. Frequently found in lawns, it tolerates shade and repeated mowing. This grass is a cause of hay fever in May and June.

Top left: Flowering
Above: Collar and ligule
Bottom left: Blade and ligule

Orchardgrass

Saltgrass

Distichlis spicata

HEIGHT: 5–16 inches

GROWTH: Native perennial, warm-season

HABITAT: Saline or alkaline marshes

DESCRIPTION: Saltgrass forms large colonies from white, creeping rhizomes. Stiff leaves are conspicuously two-ranked, projecting alternately on opposite sides of the stem. Blades are folded or inrolled with sharply pointed tips. Male and female flowers, similar in appearance, are on different plants (dioecious) in strongly compressed, awnless spikelets.

COMMENTS: The name *Distichlis* means "two rows," referring to the blade arrangement. Glands on the leaves exude salt, an adaptation that allows the species to utilize salty water. Seeds, young plants, and rootstocks are used as food by the Canada goose.

Saltgrass

Saltgrass

Male inflorescence

Two-ranked leaves

Blade and short ligule

Barnyardgrass

Barnyardgrass

Echinochloa muricata

HEIGHT: 1–5 feet

GROWTH: Native annual, warm-season

HABITAT: Moist to wet soil in open prairie draws, shorelines, mud flats, and disturbed areas

DESCRIPTION: This coarse plant can be branched or sprawling at the base. The sheaths are flattened. The leaf blades are hairless with wavy margins; each has a conspicuous midvein. Barnyardgrass is unusual because it lacks ligules. Large, one-flowered spikelets bear up to ⅝-inch-long awns. Seed heads are often reddish to dark purple.

COMMENTS: Seeds are an important food for a variety of ducks and other birds. A very similar nonnative species, *Echinochloa crusgalli*, differs by having a line of hairs at the base of the lemma tip (visible only with magnification).

No ligule

Flattened sheath

Barnyardgrass

Blade

Flowering

Spikelets

Group 5

Heads broad and open, oval or pyramidal in outline (appearing diffuse and airy)

ticklegrass (*Agrostis hyemalis*)
prairie threeawn (*Aristida oligantha*)
smooth brome (*Bromus inermis*)
prairie sandreed (*Calamovilfa longifolia*)
Scribner's panicum (*Dichanthelium oligosanthes*)
fall witchgrass (*Digitaria cognata*)
Carolina lovegrass (*Eragrostis pectinacea*)
purple lovegrass (*Eragrostis spectabilis*)
sand lovegrass (*Eragrostis trichodes*)
rice cutgrass (*Leersia oryzoides*)
scratchgrass (*Muhlenbergia asperifolia*)
witchgrass panicum (*Panicum capillare*)
fall panicgrass (*Panicum dichotomiflorum*)
switchgrass (*Panicum virgatum*)
Kentucky bluegrass (*Poa pratensis*)
tall fescue (*Schedonorus arundinaceus*)
Johnsongrass (*Sorghum halepense*)
sand dropseed (*Sporobolus cryptandrus*)
purpletop tridens (*Tridens flavus*)

Ticklegrass

Spikelets

Ticklegrass

Agrostis hyemalis

HEIGHT: 8–24 inches

GROWTH: Native perennial, cool-season

HABITAT: Mesic to dry open ground, upland prairies

DESCRIPTION: Stems of ticklegrass are slender and erect, forming delicate tufts. Leaf sheaths and blades are hairless. Blade margins are often inrolled. The branches of the flower heads are nodding, fine, and hairlike, giving it the name "hairgrass." The one-flowered spikelets are at the branch tips.

COMMENTS: Roadside and fiery skipper larvae feed on this grass. Several species of *Agrostis* are cultivated for lawns and golf courses.

Plants turn white in late June

Ticklegrass

Another view of
spikelets

Leaf blade

Prairie Threeawn

Aristida oligantha

HEIGHT: 1–2 feet

GROWTH: Native annual, warm-season

HABITAT: Dry or disturbed soil in pastures, roadsides, and fallow fields

DESCRIPTION: Stems are wiry and many-branched. The leaf blades are smooth and often inrolled. The 1-inch-long spikelets are narrow, one-flowered, and topped by three-branched, 1- to 2½-inch-long awns that bend sharply near the bases.

COMMENTS: The long spikelet and awns can cause injury to grazing animals. Prairie threeawn has a tendency to thrive in poor soil. Plants appear whitish when seeds ripen. There are seven species of threeawn in the region (250 worldwide).

Prairie Threeawn

Spikelets

Prairie Threeawn

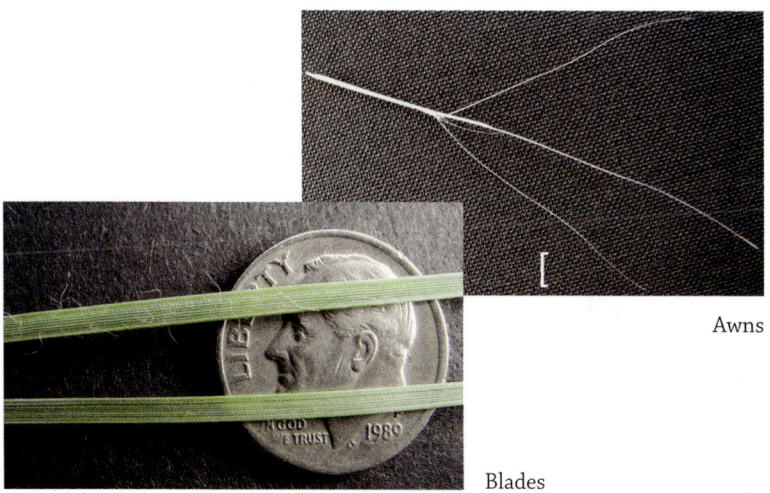

Awns

Blades

SIMILAR SPECIES: Purple threeawn (*Aristida purpurea*) is a native, perennial grass found in dry prairies in the western ⅔ of the three-state area in sandy or gravelly soils. It is most common in the areas with 15 to 25 inches of annual precipitation. It differs from prairie threeawn by having broad and unequal glumes. Also, the inflorescence has a purplish-red color.

Purple threeawn

A. purpurea spikelets

Smooth Brome

Smooth Brome

Bromus inermis

HEIGHT: 16–45 inches

GROWTH: Perennial introduced from
Europe, China, and Siberia, cool-season

HABITAT: Roadsides, pastures, and waste
places

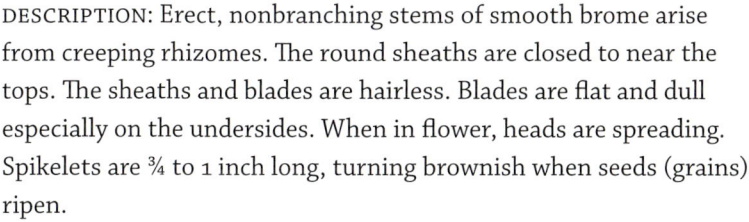

DESCRIPTION: Erect, nonbranching stems of smooth brome arise
from creeping rhizomes. The round sheaths are closed to near the
tops. The sheaths and blades are hairless. Blades are flat and dull
especially on the undersides. When in flower, heads are spreading.
Spikelets are ¾ to 1 inch long, turning brownish when seeds (grains)
ripen.

COMMENTS: This long-lived grass is widely cultivated for pasture and
hay production. It is important as forage for early and late season
grazing, but fertilization is recommended. It invades native prairie;
due to its aggressiveness, it can be considered weedy. It is common
along roadsides.

Flowering inflorescence

Smooth Brome

Blade

Closed sheath

Rhizomes

Prairie Sandreed

Calamovilfa longifolia

HEIGHT: 20–70 inches

GROWTH: Native perennial, warm-season

HABITAT: In a variety of sandy soils, often dominant in sandhill prairies

DESCRIPTION: Erect stems are mostly soli-tary, forming large, open colonies from strong rhizomes. The sheaths are veined and have short hairs on the margins near the tops. The stiff, smooth blades are flat at the bases and inrolled at the tips. The flower heads are loosely spreading. Spikelets are pale and flattened. Long, straight, white hairs can be found at the base of each floret.

COMMENTS: The plant is pale green, later becoming a straw yellow color. Prairie sandreed is tolerant of dry conditions and binds loose soil. It is common and widespread in the Nebraska Sandhills.

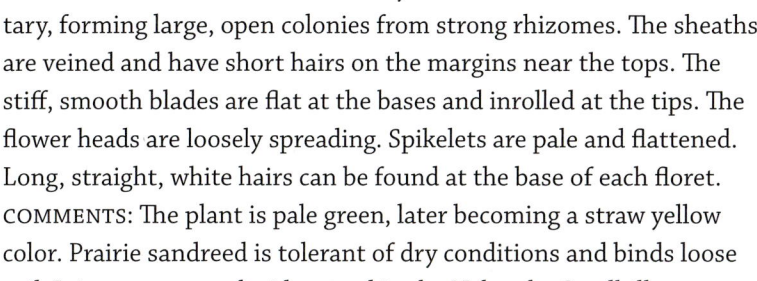

Collar region

Leaf blade

Prairie Sandreed

Flowering spikelets

Open floret showing white hairs

Scribner's Panicum

Scribner's Panicum

Dichanthelium oligosanthes

HEIGHT: 6–24 inches

GROWTH: Native perennial, cool-season

HABITAT: Upland prairies

DESCRIPTION: These tufted plants lack rhizomes. Stems are often more spreading than erect. The sheaths are usually hairy. Short, wide blades are relatively thick and stiff, each tapering quickly to a point.

COMMENTS: *Dichanthelium* species have two overlapping flower periods: Scribner's panicum blooming in the spring (May–July) and again in the fall (June–September). The short, ground-hugging leaves at the base of this and other species of *Dichanthelium* will often remain green all winter. There are at least 10 species of *Dichanthelium* growing in the three-state area.

Inflorescence

Leaf blade

Scribner's Panicum

Collar and hairy sheath

Green leaves in January

Fall Witchgrass

Digitaria cognata

HEIGHT: 8–28 inches

GROWTH: Native perennial, warm-season

HABITAT: Dry soil of upland prairies and sand hills

DESCRIPTION: Fall witchgrass has erect to sprawling stems. The sheaths are sparsely to densely hairy. What might appear to be auricles are thickened edges of the sheaths. Blade margins are wavy. The slender branches of the flower head are slightly wavy.

COMMENTS: At maturity, masses of these flower heads have a purplish-brown color that is recognizable at a distance. This trait is similar to that seen in purple lovegrass.

Spikelets

Fall Witchgrass

Fall Witchgrass

Leaf blade

Collar region

Hairs at the rachis axis

Opposite, below:
Fall witchgrass
in September

Carolina Lovegrass

Spikelets

Carolina Lovegrass

Eragrostis pectinacea
HEIGHT: 4–23 inches
GROWTH: Native annual, warm-season
HABITAT: Open, moist ground in old fields,
roadsides, and waste places
DESCRIPTION: These are densely tufted and
branching plants. The upper collar margins of each sheath have a tuft
of hairs. Blades are narrow and up to 6 inches long. The flower heads
are ovate to triangular in outline. Spikelets tend to lie parallel to the
branches.
COMMENTS: Oblong grains (seeds) are yellowish to reddish brown,
slightly flattened, and less than 1/16 inch long. This species is very
common along gravel road edges.

Collar

Blade

Purple Lovegrass

Hairs at the main axes of the inflorescence

Purple Lovegrass

Eragrostis spectabilis

HEIGHT: 10–30 inches

GROWTH: Native perennial, warm-season

HABITAT: Dry soil of upland prairies

DESCRIPTION: Tufts of stems arise from
knotty bases. The sheaths are flattened with
long, white hairs at the collars. Leaves are relatively few. Flowering
branches are stiff and spreading with short tufts of white hair at the
junctions of the main branches. The spikelets are strongly flattened
and purplish.

COMMENTS: Nestled low among the other grasses, this lovegrass
forms scattered masses of hazy reddish-purple color. The scientific
name *spectabilis* means showy. The airy seed heads add beauty to
dried bouquets.

Blade

Spikelets

Hairs at leaf collar, leaf involute

Purple Lovegrass

Reddish color

Short rhizomes (budding) and long, thick roots

Sand Lovegrass

Eragrostis trichodes

HEIGHT: 2–5 feet

GROWTH: Native perennial, warm-season

HABITAT: Upland prairies, sandy soils

DESCRIPTION: Sand lovegrass is tufted and
erect. The sheaths have prominent hairs on

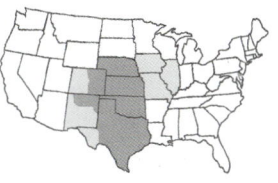

the upper margins and the collars. Large flower heads are ovate in
outline and have the appearance of clouds of mist along roadsides.

COMMENTS: Sand lovegrass makes excellent livestock forage and will
decrease when overgrazed. The genus name derives from Eros, the
God of Love, and *agrostis*, meaning "forage" or "field grass." Some
say the name is used because there are species in which the spikelets
appear heart-shaped.

Spikelets

Sand Lovegrass

Blade

Collar with hair tufts

Rice Cutgrass

Rice Cutgrass

Leersia oryzoides

HEIGHT: 20–50 inches

GROWTH: Native perennial, cool-season

HABITAT: Wet soil near springs, marshes, and along stream banks

DESCRIPTION: Rice cutgrass forms colonies from rhizomes. The foliage is a light yellowish-green color. The stems are weak and sprawling. Swollen nodes are covered by distinctive short, white hairs. The sheaths are covered with downward-pointing short, rough hairs. The blades are flat and also covered with short, spinelike hairs, especially on the margins. These hairs cause the plant to cling and can cut skin. The spikelets are strongly flattened, overlapping, and oriented along one side of the flower branches.

COMMENTS: The least skipper and Peck's skipper larvae feed on the foliage. Ducks not only eat the seeds, but will pull up and eat the roots.

Spikelets flowering

Rice Cutgrass

Leaf blade

Hairy nodes and sheath is shorter than internode

Scratchgrass

Muhlenbergia asperifolia
HEIGHT: 4–18 inches
GROWTH: Native perennial, warm-season
HABITAT: Low, moist soil, often alkaline
DESCRIPTION: Stems of scratchgrass arise
from long, slender rhizomes. Branching

from the base, the stems are slender, smooth, and shiny. The sheaths
are somewhat flattened (keeled). The blades are flat or inrolled near
the tips and have rough margins. The flower heads are open and
ovate in outline.

COMMENTS: The grains of scratchgrass may be infected with a smut
fungus, *Tilletia*, causing them to enlarge and turn black. Twelve other
species of *Muhlenbergia* occur in this region.

Scratchgrass

Scratchgrass

Leaf blade

Spikelets flowering

Two-ranked leaves

Witchgrass Panicum

Witchgrass Panicum

Panicum capillare

HEIGHT: 8–28 inches

GROWTH: Native annual, warm-season

HABITAT: Mesic to dry and disturbed areas, fallow fields, gravel bars, often sandy soils

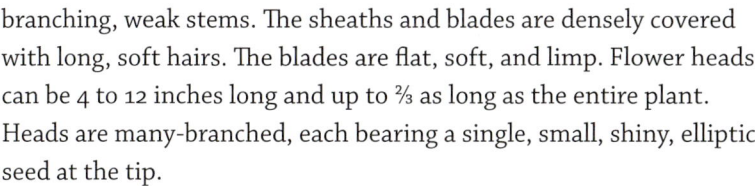

DESCRIPTION: Witchgrass panicum has branching, weak stems. The sheaths and blades are densely covered with long, soft hairs. The blades are flat, soft, and limp. Flower heads can be 4 to 12 inches long and up to ⅔ as long as the entire plant. Heads are many-branched, each bearing a single, small, shiny, elliptic seed at the tip.

COMMENTS: Turning purplish at maturity, the seed head breaks away and tumbles in the wind, scattering seed. The seeds are a source of food for many birds. This is a weedy species of little forage value for livestock.

Witchgrass Panicum

Hairy leaf blade

Spikelets

Softly hairy sheath

Fall Panicgrass

Panicum dichotomiflorum

HEIGHT: 15–40 inches

GROWTH: Native annual, warm-season

HABITAT: Open, disturbed soil in pastures and along roadsides

DESCRIPTION: Stems of fall panicgrass are spreading with erect tips. The stems are slightly compressed and branching, and lower nodes are swollen. Each leaf blade has a white midrib and rough margins. The distinctive lower glume is broad, short, and obtuse.

COMMENTS: This is a weedy grass that is not highly palatable to livestock. However, larvae of the tawny-edged skipper feed on this grass. The ovoid, glossy seeds of *Panicum* species are important as food to a large variety of birds. Foliage is eaten by fur and game mammals.

Fall Panicgrass

Annual grass roots are fibrous, shallow, and without rhizomes

Fall Panicgrass

Blade

Open sheath

Spikelets flowering

Switchgrass

Switchgrass

Panicum virgatum

HEIGHT: 2–7 feet

GROWTH: Native perennial, warm-season

HABITAT: Prairie, especially moist lowland

DESCRIPTION: Arising from tough rhizomes, switchgrass stems are often clustered in

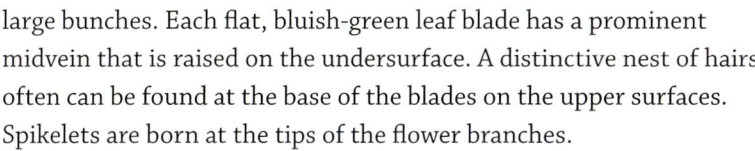

large bunches. Each flat, bluish-green leaf blade has a prominent midvein that is raised on the undersurface. A distinctive nest of hairs often can be found at the base of the blades on the upper surfaces. Spikelets are born at the tips of the flower branches.

COMMENTS: Switchgrass makes excellent nesting cover, and seeds are used by birds and small mammals. Leaves are food for larvae of the Delaware skipper. In early stages of growth, switchgrass is an excellent pasture grass. Foliage becomes a flaming orange color in the fall. There are many cultivated varieties of switchgrass grown as garden ornamentals.

Switchgrass

Leaf blade

Hairs at the leaf base

Spikelets flowering

Kentucky Bluegrass

Poa pratensis
HEIGHT: 4–40 inches
GROWTH: Perennial introduced from
Europe, cool-season
HABITAT: Fields, woods, and pastures
DESCRIPTION: Kentucky bluegrass is sod

forming from rhizomes. The sheaths are closed to about ½ their lengths. The flower heads are pyramid-shaped when in bloom. Spikelets are flattened. The two major identifying features of this grass are the boat-shaped tips of the leaf blades and the silky, cob-webby hairs at the bases of the florets.

COMMENTS: Pollen of Kentucky bluegrass is a cause of hay fever in May and June. The species is often cultivated as a lawn grass but is not drought-tolerant. Kentucky bluegrass is regarded as an invader in native prairie. It increases rapidly in overgrazed pastures but can be effectively reduced by prescribed spring burning.

Kentucky Bluegrass

Kentucky Bluegrass

Prow-shaped leaf tips

Blade and sheath

Inflorescence

Cobweb hairs

Tall Fescue

Tall Fescue

Schedonorus arundinaceus

HEIGHT: 15–42 inches

GROWTH: Perennial introduced from Eurasia, cool-season

HABITAT: Pastures and roadsides

DESCRIPTION: Stout, erect stems are tufted. Dark green blades are ridged on top, lighter color, smooth, and shiny beneath. Auricles and collars have a few short hairs on the margins. The inflorescences are spread during flowering and are somewhat nodding.

COMMENTS: Tall fescue is grown as forage and lawn turf. It can easily be confused with meadow fescue (*Schedonorus pratensis*), once popular as a pasture grass. Tall fescue is frequently infected by systemic fungus (*Neotyphodium*), which creates toxicity problems for grazing animals. The fungus benefits the plant, conferring insect and drought resistance.

Short hairs on the auricles

Blade

Tall Fescue

Above: Blooming florets

Above right: Inflorescence contracts as seeds mature

Mature plants in autumn after seeds have fallen

Johnsongrass

Sorghum halepense

HEIGHT: 2–7 feet

GROWTH: Perennial introduced from Mediterranean region, Africa, and India, warm-season

HABITAT: Old fields, roadsides, and waste places, often on moist soils

DESCRIPTION: Extensive thick, creeping rhizomes enable Johnsongrass to quickly form colonies. Blades are flat and wide, each with a thickened, whitish midvein. Smooth, shiny seeds are reddish to purple, turning straw color with age.

COMMENTS: Johnsongrass is frequently planted as hay or a forage crop. Stressed plants produce prussic acid, making them poisonous to livestock. Johnsongrass invades crop fields and roadsides. It is so difficult to control that it is listed as legally noxious in Kansas. It is closely related to cultivated and weedy sorghums. Well established in the south, this grass is steadily moving northward.

Collar and ligule

Leaf blade

Johnsongrass

Flowering

Rhizome

Sand Dropseed

Sand Dropseed

Sporobolus cryptandrus
HEIGHT: 12–40 inches
GROWTH: Native perennial, warm-season
HABITAT: Dry, often sandy soils
DESCRIPTION: Stems of sand dropseed
are erect and flattened or grooved on one
side. The sheaths bear a dense tuft of white hairs at the collars and
can be hairy on the margins. The blades are flat or inrolled toward
the pointed tips. Widely spaced flower branches spread at maturity,
forming a pyramid shape.

COMMENTS: In winter, wind-whipped leaves shred into frayed threads.
This grass has a fair forage value; however, plants increase with heavy
grazing. Several thousand small seeds can be produced in a single
head of sand dropseed. The seeds are important to ground-feeding
birds and small mammals. Plains Indians prepared the dried grains of
this species as food by grinding and making them into mush.

Top left: Blade
Above: Inflorescence pushing out
of the sheath
Left: Sheath margin is not always
this hairy

Purpletop Tridens

Purpletop Tridens

Tridens flavus
HEIGHT: 2–5 feet
GROWTH: Native perennial, warm-season
HABITAT: Mesic conditions in prairies and open woods

DESCRIPTION: The sheaths of purpletop tridens are flattened at the bases with dense hairs at the collars. Blades are flat and smooth. Open flower heads have drooping branches and are purplish in color, becoming tan.

COMMENTS: The common name "greasegrass" originates from an oily, sticky substance produced on the stems and other parts of this species. Purpletop tridens has a distinctive odor possibly associated with this oily coating. The larvae of crossline, Leonard's, and zabulon skippers, as well as the little glassywing and common wood nymph, use this grass as food.

Purpletop Tridens

Above: Ligule of short hairs

Right: Leaf blade

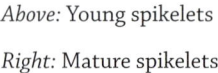

Above: Young spikelets

Right: Mature spikelets

Group 6

Heads of one to several short to long branches spaced along the stem, most perpendicular to the stem (appearing comblike or ladderlike)

sideoats grama (*Bouteloua curtipendula*)
buffalograss (*Bouteloua dactyloides*)
blue grama (*Bouteloua gracilis*)
hairy grama (*Bouteloua hirsuta*)
sand paspalum (*Paspalum setaceum*)
tumblegrass (*Schedonnardus paniculatus*)
prairie cordgrass (*Spartina pectinata*)

Sideoats Grama

Sideoats Grama

Bouteloua curtipendula
HEIGHT: 10–40 inches
GROWTH: Native perennial, warm-season
HABITAT: Plains, upland prairies, rocky hillsides

DESCRIPTION: Tufted plants arise from short rhizomes. The blade margins have perpendicular, widely spaced hairs with enlarged pustular bases. The many short flower branches hang mostly on one side of the stems.

COMMENTS: This grass is widespread but is most important in mixedgrass prairies, providing excellent forage for many kinds of animals. In the fall, leaves of the plant base become reddish. The bare, pale yellow flower stalks remain standing all winter. Sideoats grama is commonly used in restoration seed mixes and is easy to establish.

Sideoats Grama

Blade with hairs

Red anthers

Buffalograss

Bouteloua dactyloides

HEIGHT: 2–8 inches

GROWTH: Native perennial, warm-season

HABITAT: Plains and dry prairies, uncommon in sandy soils

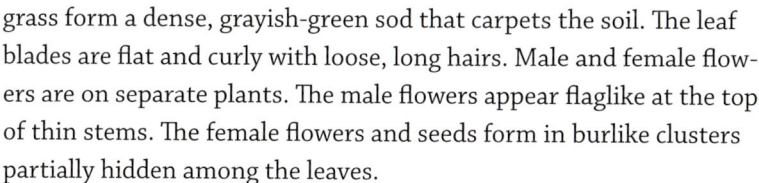

DESCRIPTION: Creeping stolons of buffalograss form a dense, grayish-green sod that carpets the soil. The leaf blades are flat and curly with loose, long hairs. Male and female flowers are on separate plants. The male flowers appear flaglike at the top of thin stems. The female flowers and seeds form in burlike clusters partially hidden among the leaves.

COMMENTS: Still listed as *Buchloe dactyloides* in many manuals, the name was recently changed to reflect a closer relationship to the genus *Bouteloua*. Although it is found in all three types of prairies, buffalograss is characteristic of shortgrass prairies. Competition from taller grasses will shade it out. Sod homes of early settlers were often made from buffalograss. Buffalograss is drought-resistant and cures well for winter grazing.

Leaf blade

Collar and leaf arrangement

Buffalograss

Seed-bearing "burs"

Buffalograss

Male inflorescence

Female inflorescence

Stolon

Blue Grama

Blue Grama

Bouteloua gracilis

HEIGHT: 6–20 inches

GROWTH: Native perennial, warm-season

HABITAT: Plains and dry prairies

DESCRIPTION: Stems of blue grama are erect and slender. Leaf blades are narrow with long hairs at the collars. Flower branches often curve downward at the ends resembling eyebrows. Spikelets are crowded in two rows on one side of each branch (rachis), appearing comblike.

COMMENTS: Blue grama is a dominant grass in mixed and shortgrass prairies. Extremely drought-tolerant and attractive, blue grama makes a good lawn grass where frequent dry periods prevail. The name, *Bouteloua*, is in honor of the eighteenth-century Spanish botanists Claudio and Esteban Boutelou y Soldevilla.

Greenish-yellow anthers

Blade

Blue Grama

In November; note the "eyebrows"

Hairy Grama

Bouteloua hirsuta

HEIGHT: 6–28 inches

GROWTH: Native perennial, warm-season

HABITAT: Plains and rocky hills in dry, shallow soil

DESCRIPTION: This grass is tufted with leaves mostly at the base of the plant. The sheaths have long hairs at the collars. Lower margins of the blades possess hairs with pustular bases. Spikelets are crowded and appear comblike. The flower branches often curve upward at the ends, resembling eyelashes. Flower branches (rachises) extend beyond the spikelets to needlelike points.

COMMENTS: The fine and highly branched roots of hairy grama penetrate the soil to a depth of 4 feet. Most common in mixed and short-grass prairies, hairy grama is a valuable species for wildlife, including pronghorn, deer, small mammals, and many birds.

Rachis projects sharply at the tip

Leaf blade

Hairy Grama

Sand Paspalum

Paspalum setaceum

HEIGHT: 6–30 inches

GROWTH: Native perennial, warm-season

HABITAT: Scattered on prairies, roadsides, and dry, sandy soils

DESCRIPTION: Sand paspalum is tufted from hardened knotty bases and bears erect to spreading stems. The sheaths may be somewhat flattened and hairy on the margins and collars. Blades are flat, soft, and thinly hairy. The flower heads have one to four narrow, widely spaced branches. The one-flowered, flattened spikelets are nearly round and are arranged in two rows on one side of each branch (rachis).

COMMENTS: Seeds of sand paspalum are food for birds and small mammals. This species is widespread and variable. Sand paspalum is scattered, never dominating the plant community.

Blade with wavy edges

Spikelets

Sand Paspalum

Sand Paspalum

SIMILAR SPECIES: Another native, hairyseed paspalum (*Paspalum pubiflorum*) is found on moist ground in most of Oklahoma and the eastern half of Kansas but is absent in Nebraska. It has more flower branches than sand paspalum and the spikelets are oval and larger than those of sand paspalum.

Hairyseed paspalum

Hairyseed paspalum

Tumblegrass

Leaf arrangement on the stem

Tumblegrass

Schedonnardus paniculatus
HEIGHT: 8–20 inches
GROWTH: Native perennial, warm-season
HABITAT: Prairies and plains
DESCRIPTION: Tumblegrass forms
short tufts. The sheaths are flattened.
Membranous ligules are up to ⅛ inch long and extend down the
sheath margins. Stiff blades are folded and crowded at the bases of
the plants. The long, arching flower stems are more than half the entire plant height. Tiny, one-flowered spikelets are tightly pressed in
two rows along the slender, wiry branches.
COMMENTS: When seeds mature, the stems break away, tumbling in
the wind. These dry, curving seed stems can work their way up pants
legs, poking and scratching, requiring dexterous manipulation to dis-
lodge them.

Leaf blade

Spikelets flowering

Prairie Cordgrass

Prairie Cordgrass

Spartina pectinata

HEIGHT: 3–7 feet

GROWTH: Native perennial, warm-season

HABITAT: Wet soil, marshes, ditches, low prairies

DESCRIPTION: This coarse grass forms colonies from stout rhizomes that bind the soil and prevent erosion. The sheaths are round. The long, flat blades each have a prominent midrib and rough margins that can cut skin. Flower branches are 1½ to 4 inches long, arranged alternately in two rows parallel to the central stems.

COMMENTS: Cordgrass once covered thousands of acres along river lowland, but now nearly all of it has been converted to cropland. Several moth species are specialist feeders of this grass. Muskrats use cordgrass for food and nesting. Cordgrass was used as thatch and building material by Ponca and Omaha tribes.

Top left: Blade

Above: Comblike flower branches

Left: Rhizomes

Prairie Cordgrass

Plants turn yellow in fall

Group 7

Heads of two to many narrow branches radiating in whorls near the top of the stem or along the stem (appearing fingerlike or spokelike, as a wheel)

big bluestem (*Andropogon gerardii*)
windmillgrass (*Chloris verticillata*)
hairy crabgrass (*Digitaria sanguinalis*)
goosegrass (*Eleusine indica*)
eastern gammagrass (*Tripsacum dactyloides*)

Big Bluestem

Big Bluestem

Andropogon gerardii

HEIGHT: 2–7 feet

GROWTH: Native perennial, warm-season

HABITAT: Prairies and plains

DESCRIPTION: Plants form large clumps
from short rhizomes. The roots can extend

12 feet deep. Stems are round, grooved on one side, and covered with
a whitish, waxy bloom (glaucous). The blades are flat, each with a
thickened midvein. Palmate flower branches are purplish to yellowish
brown with hairlike awns spirally twisted and bent near the bases.

COMMENTS: Big bluestem is a dominant component of the tallgrass
prairie vegetation and produces high-quality forage for all kinds of
livestock. Plants turn a red-orange color after frost. It was called
hade-zhide by the Omaha and Ponca Indians, meaning "red hay."

Big Bluestem

Top left: Blade

Above: Flowers

Left: The white area below the node is the flat side of the stem

SIMILAR SPECIES: Sand Bluestem (*Andropogon hallii*) is similar to big bluestem. The most striking difference is in the flower branches, which are densely hairy. Both species may be found growing together, but sand bluestem is rare on soils that are not sandy. Sand bluestem is found through all of Nebraska except the southeast corner, in the west ⅔ of Kansas, and in the west half of Oklahoma.

Sand bluestem

Windmillgrass

Chloris verticillata

HEIGHT: 4–16 inches

GROWTH: Native perennial, warm-season

HABITAT: Dry, upland prairies and lawns in a variety of soils

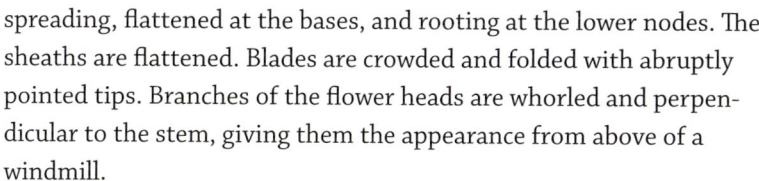

DESCRIPTION: Stems of windmillgrass are spreading, flattened at the bases, and rooting at the lower nodes. The sheaths are flattened. Blades are crowded and folded with abruptly pointed tips. Branches of the flower heads are whorled and perpendicular to the stem, giving them the appearance from above of a windmill.

COMMENTS: When mature, the seed heads break off and tumble in the wind. *Chloris* means "pale green." In Greek mythology Chloris is the goddess of flowers and of spring. There are 55 species of *Chloris*, mostly in tropical regions.

Windmillgrass

Heads break and tumble, collecting in drifts

Windmillgrass

Leaf blade

Flattened sheath and folded leaf

Spikelets

Hairy Crabgrass

Hairy Crabgrass

Digitaria sanguinalis
HEIGHT: 8–30 inches
GROWTH: Annual introduced from Eurasia, warm-season
HABITAT: Lawns, crop fields, roadsides, and pastures
DESCRIPTION: Spreading stems are freely branching and rooting at the nodes. The sheaths are keeled and covered with long, stiff hairs on the surfaces and the margins. Blades are flat, hairy on both surfaces, and each with an obvious midrib. At the top of the stems is a whorl of several fingerlike flower branches. One-seeded spikelets are arranged in two rows along one side of each branch.
COMMENTS: Hairy crabgrass is an aggressive lawn weed. It is palatable to livestock and can be grazed, but provides little forage. The ¹⁄₁₆-inch-long seeds with pale translucent margins are eaten by many birds, including wild turkeys.

Plants root where nodes touch the ground

Sheath with stiff hairs

Hairy Crabgrass

Above: Hairy leaf

Right: Flowering

SIMILAR SPECIES: Smooth crabgrass (*Digitaria ischaemum*) is usually a bit smaller than hairy crabgrass, the sheaths and blades have little or no hair, and the second glumes are nearly as long as the spikelets. *Digitaria* is a large genus, found mostly in tropical and subtropical regions, with about 220 species worldwide.

Smooth crabgrass blade

Smooth crabgrass

Goosegrass

Eleusine indica

HEIGHT: 4–30 inches

GROWTH: Annual introduced from Africa, warm-season

HABITAT: Roadsides, fields, and lawns in heavy and compacted soils

DESCRIPTION: Goosegrass is dark green and spreading or prostrate. The flattened stems and sheaths sometimes root at the lower nodes. The sheath margins and blades often have long hairs near the collars. The blades are folded or V-shaped with blunt tips. Whorled, fingerlike inflorescence branches have a silvery sheen. Spikelets are strongly flattened laterally in two rows on one side of each flattened branch (rachis).

COMMENTS: Goosegrass is a common weed in urban areas, where it can be found growing in pavement cracks.

Inflorescence

Spikelets

Goosegrass

Leaf arrangement is two-ranked
with a flattened sheath

Leaf blade

Goosegrass

SIMILAR SPECIES: Bermudagrass (*Cynodon dactylon*) has the palmately branched flower heads and crowded, flat stems as some other species in Group 7. The low-growing, mat-forming plants have both rhizomes and stolons. Bermudagrass is a nonnative perennial that is cultivated as a lawn grass, but it can be weedy. It is common in Oklahoma and Kansas but not persistent in Nebraska.

Bermudagrass

Bermudagrass leaf arrangement

Eastern Gammagrass

Eastern Gammagrass

Tripsacum dactyloides

HEIGHT: 4–8 feet

GROWTH: Native perennial, warm-season

HABITAT: Moist to mesic prairies in a wide variety of soils

DESCRIPTION: Eastern gammagrass grows in large clumps. Tops of the short, stout rhizomes sometimes can be seen by brushing away loose soil at the center of the plant. Stems are flat at the bases and purplish. The leaf blades are hairless, each with a white central rib and rough margins. The inflorescences usually have two to three branches. Male flowers occur above the female flowers on the same branches (monoecious).

COMMENTS: Gammagrass is a high producer of palatable forage and host to larvae of byssus skipper. The name *Tripsacum* means "polished grain," referring to the shiny, teardrop-shaped grains.

Leaf blade with white midrib

Short internodes with attached spikelets containing fruits (seeds)

Eastern Gammagrass

Male spikelets

Female spikelets

Short, thick rhizome

Grasslike Plants

Some plants look similar to grasses—sedge, rush, cattail, horsetail, blue-eyed-grass, death camas, spiderwort, and wild onion at certain stages might be mistaken for a grass. Some of these plants are closely related to grasses. Plants of the sedge (Cyperaceae) and rush (Juncaceae) families resemble grasses so closely that they are often described with grasses.

Sedges are a relatively large family. They are widespread and frequently prefer moist habitats. Sedges can be distinguished from grasses by their usually solid, triangular stem, three-ranked leaves, closed (fused) leaf sheath around the stem, and lack of nodes. Many sedges (as in *Carex meadii*) have clearly different male and female flowers grouped in separate spikes on a single plant. In ancient Egypt, papyrus paper was made from the pithy stems of a plant in the sedge family, *Cyperus papyrus*. Tubers of a cultivated variety of yellow nutsedge, *Cyperus esculentus*, called "chufa" or "earth almonds," are still eaten raw or cooked just as they were 5,000 years ago in Egypt. Other groups of plants belonging to the sedge family, bulrush (*Schoenoplectus*, *Scirpus*) and spikerush (*Eleocharis*), were much used in basketry by North American Indians.

Fragrant flatsedge (*Cyperus odoratus*)

Heavy sedge (*Carex gravida*)

Woolly sedge
(*Carex pellita*)

Mead's sedge (*Carex meadii*)

Mead's sedge

Yellow nutsedge tubers

Yellow nutsedge (*Cyperus esculentus*)

Chairmaker's bulrush (*Schoenoplectus pungens*)

Cloaked bulrush (*Scirpus pallidus*)

The triangular stem
of cloaked bulrush
in cross-section

Nodding bulrush (*Scirpus pendulus*)

Pale spikerush (*Eleocharis macrostachya*)

The rush family, Juncaceae, has fewer species. They typically inhabit wet places. Like grasses and sedges, rushes are worldwide in distribution and are wind-pollinated. Rush flowers are rarely unisexual, as opposed to sedges, and differ from grass flowers in that each flower has six tan or brown tepals (the term for nearly identical petals and sepals). Rush stems are round and lack nodes. Leaves are small, few, and sometimes tubular. Mature seed capsules are critical to species identification. The seeds and roots of these plants provide food for a variety of wildlife.

Only a few examples of the many species of sedges and rushes are pictured here.

Inland rush (*Juncus interior*)

Torrey's rush (*Juncus torreyi*)

Slimpod rush (*Juncus diffusissimus*)

Advanced Agrostology—Looking Deeper

Agrostology is the study of grasses. Although this book is intended to be used without the need of magnification, eventually every agrostologist must resort to magnification to see small structures used in identification. An interesting structure to look for, though not often required for identification, is the lodicule.

The lodicules, located inside the lemma and palea at the base of the ovary, are thought to be remnants of flower petals, which are otherwise lacking in grasses. Lodicules serve a significant function in the development of the grass flower. Prior to pollination, the lodicules swell, forcing the lemma and palea apart and exposing the flower for pollination.

To find this tiny, hidden structure, one needs a microscope, tweezers, and straight pins to open the floret. Bring in several different types of grass spikelets, and see if you can find the lodicules in each of them. Be sure to collect the spikelets near flowering time. With tweezers, carefully remove a single floret from the spikelet, and use pins to help pull apart the outer bracts (palea and lemma). The lodicules will look something like those in the illustration below. In *Dactylis*, *Hordeum*, or *Koeleria*, lodicules are a bit larger and may be easier to find.

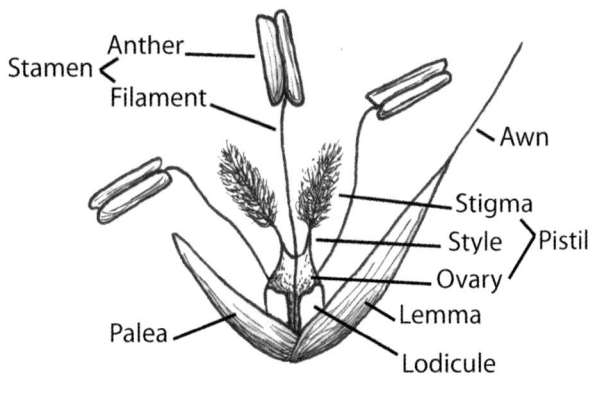

OPEN FLORET

Flowering Dates

The grasses are listed below first by the dates they may typically be found flowering (from earliest to latest) and then alphabetically by scientific name.

Grass	April	May	June	July	August	Sept	Oct
Downy brome *Bromus tectorum*	████████████████						
Kentucky bluegrass *Poa pratensis*	████████████████						
Prairie wedgegrass *Sphenopholis obtusata*	███████████████████████						
Jointed goatgrass *Aegilops cylindrica*		██████████					
Ticklegrass *Agrostis hyemalis*		██████████					
Smooth brome *Bromus inermis*		██████████					
Orchardgrass *Dactylis glomerata*		██████████					
Little barley *Hordeum pusillum*		██████████					
Prairie Junegrass *Koeleria macrantha*		██████████					
Sixweeks fescue *Vulpia octoflora*		██████████					
Japanese brome *Bromus japonicus*		█████████████████					
Rye brome *Bromus secalinus*		█████████████████					
Virginia wildrye *Elymus virginicus*		█████████████████					
Fowl mannagrass *Glyceria striata*		█████████████████					
Needle-and-thread *Hesperostipa comata*		█████████████████					

Grass	April	May	June	July	August	Sept	Oct
Porcupinegrass *Hesperostipa spartea*		▬	▬	▬	▬		
Purple threeawn *Aristida purpurea*		▬	▬	▬	▬	▬	
Buffalograss *Bouteloua dactyloides*		▬	▬	▬	▬	▬	▬
Reed canarygrass *Phalaris arundinacea*		▬	▬	▬	▬	▬	
Scribner's panicum *Dichanthelium oligosanthes*			▬	▬	▬	▬	▬
Scratchgrass *Muhlenbergia asperifolia*		▬	▬	▬	▬	▬	▬
Tall fescue *Schedonorus arundinaceus*		▬	▬	▬	▬	▬	▬
Canada wildrye *Elymus canadensis*			▬	▬	▬		
Western wheatgrass *Pascopyrum smithii*			▬	▬	▬		
Eastern gammagrass *Tripsacum dactyloides*			▬	▬	▬		
Foxtail barley *Hordeum jubatum*			▬	▬	▬		
Sand paspalum *Paspalum setaceum*			▬	▬	▬	▬	
Tumblegrass *Schedonnardus paniculatus*			▬	▬	▬	▬	
Hairy grama *Bouteloua hirsuta*			▬	▬	▬	▬	▬
Prairie sandreed *Calamovilfa longifolia*			▬	▬	▬	▬	▬
Windmillgrass *Chloris verticillata*			▬	▬	▬	▬	▬
Saltgrass *Distichlis spicata*			▬	▬	▬	▬	▬
Barnyardgrass *Echinochloa muricata*			▬	▬	▬	▬	▬
Fall panicgrass *Panicum dichotomiflorum*			▬	▬	▬	▬	▬

Grass	April	May	June	July	August	Sept	Oct
Green bristlegrass *Setaria viridis*			▬	▬	▬	▬	
Bermudagrass *Cynodon dactylon*			▬	▬	▬	▬	▬
Stinkgrass *Eragrostis cilianensis*			▬	▬	▬	▬	▬
Johnsongrass *Sorghum halepense*			▬	▬	▬	▬	▬
Sand dropseed *Sporobolus cryptandrus*				▬	▬		
Caucasian bluestem *Bothriochloa bladhii*				▬	▬	▬	
Yellow bluestem *Bothriochloa ischaemum*				▬	▬	▬	
Silver beardgrass *Bothriochloa laguroides*				▬	▬	▬	
Sideoats grama *Bouteloua curtipendula*				▬	▬	▬	
Blue grama *Bouteloua gracilis*				▬	▬	▬	
Sandbur *Cenchrus longispinus*				▬	▬	▬	
Fall witchgrass *Digitaria cognata*				▬	▬	▬	
Prairie cupgrass *Eriochloa contracta*				▬	▬	▬	
Switchgrass *Panicum virgatum*				▬	▬	▬	
Yellow foxtail *Setaria pumila*				▬	▬	▬	
Prairie cordgrass *Spartina pectinata*				▬	▬	▬	
Purpletop tridens *Tridens flavus*				▬	▬	▬	
Big bluestem *Andropogon gerardii*				▬	▬	▬	
Sand bluestem *Andropogon hallii*				▬	▬	▬	

Grass	April	May	June	July	August	Sept	Oct
Goosegrass *Eleusine indica*				███	███	███	███
Carolina lovegrass *Eragrostis pectinacea*				███	███	███	███
Witchgrass panicum *Panicum capillare*				███	███	███	███
Hairyseed paspalum *Paspalum pubiflorum*				███	███	███	███
Little bluestem *Schizachyrium scoparium*				███	███	███	███
Purple lovegrass *Eragrostis spectabilis*					███	███	
Broomsedge bluestem *Andropogon virginicus*					███	███	███
Prairie threeawn *Aristida oligantha*					███	███	███
Smooth crabgrass *Digitaria ischaemum*					███	███	███
Hairy crabgrass *Digitaria sanguinalis*					███	███	███
Sand lovegrass *Eragrostis trichodes*					███	███	███
Rice cutgrass *Leersia oryzoides*					███	███	███
Rough dropseed *Sporobolus compositus*					███	███	███
Indiangrass *Sorghastrum nutans*					███	███	███

Grass Leaf Comparisons

The upper surface of each leaf blade is pictured here actual size. To find the correct leaf group, it will help to select a larger leaf on the plant. Look closely at the midrib or lack of one; at the margins for waviness, color, or hairs; and at the leaf surface for roughness, grooves, and ridges, or hairs. Some leaves are glaucous or have a distinctive color. To verify your determination, check the main descriptions. Other sources of information are listed in the references.

Leaf blades ⅛ inch wide or less

Ticklegrass

Prairie threeawn

Buffalograss

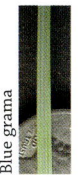
Blue grama

Hairy grama

Scratchgrass

Tumblegrass

Sixweeks fescue

Leaf blades ¼ inch wide or less

Jointed goatgrass

Caucasian bluestem

Sideoats grama

Downy brome

Windmillgrass

Fall witchgrass

Saltgrass

Goosegrass

Stinkgrass

Carolina lovegrass

Sand lovegrass

Fowl mannagrass

Porcupinegrass

Foxtail barley

Little barley

Prairie Junegrass

Western wheatgrass

Kentucky bluegrass

Little bluestem

Prairie wedgegrass

Rough dropseed

Sand dropseed

Leaf blades ⅜ inch wide or less

Silver beardgrass

Sandbur

Orchardgrass

Smooth crabgrass

Hairy crabgrass

Purple lovegrass

Tall fescue

Big bluestem

Leaf blades ½ inch wide or less

Smooth brome

Scribner's panicum

Canada wildrye

Prairie cupgrass

Rice cutgrass

Switchgrass

Fall panicgrass

Green bristlegrass

Indiangrass

Prairie cordgrass

Purpletop tridens

Sand paspalum

Yellow foxtail

Leaf blades 1 inch wide or less

Barnyardgrass

Witchgrass panicum

Prairie sandreed

Reed canarygrass

Eastern gammagrass

Leaf blades 1½ inch wide or less

Johnsongrass

Glossary

The numbers in parenthesis correspond with the illustrations.

Annual. A plant that lives only one year.

Anther. (1) Pollen-bearing structure of the flower (see stamen).

Auricle. (2) Clasping lobes on the margins of a grass leaf at the juncture of the blade and sheath.

1. Big bluestem

2. Western wheatgrass

Awn. (3) A terminal, stiff, slender, bristlelike appendage.

Axil. The upper angle formed where a stem and any part arising from it join.

Bracts. (4) Reduced leaves at the base of or surrounding the flowers.

Collar. (5) The area on the

3. Barnyardgrass

outside of a grass leaf at the juncture of the blade and sheath.

Cool-season. Grasses that grow best when soil temperatures are between 40 and 65 degrees F; also called C-3 grasses in reference to the three-carbon molecule stage during photosynthesis.

4a. Prairie sandreed

4b. Purpletop tridens; pairs of outer bracts (glumes) after seeds have fallen

5. Sand dropseed

Dioecious. Having male and female flowers on different plants.

Dominant. Plants that outcompete others by exerting the strongest ecologic influence, becoming more plentiful.

Ergot. A fungal disease of grasses in which the grain becomes enlarged, hard, and black. It can be poisonous to people and livestock.

Filament. The threadlike stalk that supports the anther (see stamen).

Floret. (6) A single flower, consisting of lemma, palea, stamens, and pistil (can also be only staminate or only pistillate).

Genus. A classification based on relationships; consisting of a collection of species.

Glade. A grassy opening in a wood or forest.

Glaucous. (7) Covered with a whitish or bluish waxy coating that rubs off easily, as with grapes and plums.

6a. Sandbur

6b. Floret opened

7. Silver beardgrass stem

Inflorescence. (8) The flowering part of a plant; the entire flower head including spikelets.

8. Prairie cordgrass

Internode. The part of a stem between two nodes.

Introduced. A plant or animal that is living in an area outside of its native range.

Invasive. Dominant colonization of a particular habitat or wild area due to loss of natural controls.

Involute. (9) Rolled inward from the upper edges.

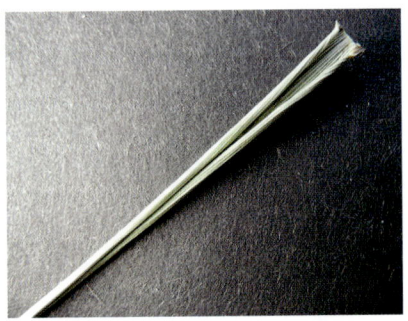

9. Canada wildrye leaf

10. Orchardgrass

Ligule. (10) The thin, collarlike appendage (membrane or line of hairs) on the inside of a grass leaf at the juncture of the blade and sheath.

Lobed. (11) In the inflorescence, unevenly clustered segments, each segment somewhat rounded in appearance.

Monoecious. Having male and female flowers at different locations on the same plant.

Native. Indigenous to an area, not imported or exotic.

Node. (12) The joint on a stem where leaves and branches originate.

Noxious. A legal designation for a plant of foreign origin that is considered injurious to agricultural interests or the public health.

11. Sandbur

12. Japanese brome

Obtuse. Dull, not pointed; denoting an angle greater than a right angle.

Palmate. (13) Radiating from a common point, like the fingers of a hand.

Perennial. A plant that lives for three or more years.

Physiographic region. The physical geography of a region.

Pistil. The female reproductive organ of a flower, consisting of a stigma, style, and ovary.

Pistillate. Applied to flowers or plants bearing pistils only (female).

Pollen. (14) Powdery contents of the anther.

Prow-shaped. (15) Curving to a point as the bow of a ship.

Pustular. A blisterlike elevation.

Rachis. (16) The main support in an inflorescence.

Rhizome. (17) A horizontal underground stem.

Savanna. A grassland community within which scattered trees occur.

13. Eastern gammagrass

14. Prairie cordgrass, each blue mark equals 1/16 inch

15. Kentucky bluegrass

16. Switchgrass

17. Switchgrass

18a. Sand lovegrass; two glumes at the base and six to eight florets

18b. Indiangrass; contains only one floret with a bent awn

Sepal. One of several leafy bracts (usually green) at the base of a flower.

Sessile. Attached directly without a supporting stalk.

Specific epithet. One element in the scientific name of a plant; it follows the genus.

Spikelet. (18) The unit of a grass inflorescence consisting of two glumes and one or more florets.

Stamen. The male reproductive organ of a flower, consisting of an anther and filament.

Staminate. Applied to flowers or plants bearing stamens only (male).

Stigma. (19) The portion of the flower (see pistil) that receives the pollen, usually feathery in grasses.

Stolon. (20) A horizontal stem trailing aboveground that roots at the nodes.

Taxonomic. As in taxonomy, the formal classification of organisms according to natural relationship.

Tufted. Clustered or bunched.

Two-ranked. (21) Each leaf along the stem is turned 108 degrees from the one before it.

Volunteer. A plant normally found only in cultivation, growing where it was not planted, and not persisting for long in the wild.

Warm-season. Grasses that begin growth when soil temperatures reach 60–65 degrees F; also called C-4 grasses in reference to the

19. Eastern gammagrass

20. Buffalograss

21. Goosegrass

four-carbon molecule stage
during photosynthesis.

Whorled. (22) A ringlike
arrangement of similar parts
arising from a common point.

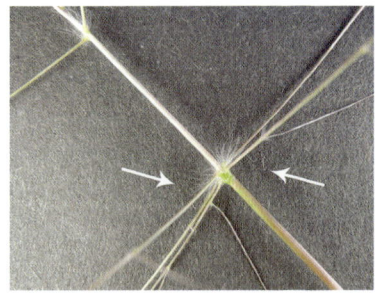

22. Fall witchgrass (central stem of the
inflorescence)

Bibliography

Allen, Durward L. *The Life of Prairies and Plains*. New York: McGraw-Hill, 1967.

Baker, Mary F. *The Book of Grasses*. Garden City, N.Y.: Doubleday, Page, 1912, http://www.archive.org/details/bookgrassesaniloobakegoog.

Barkley, Theodore M. *Field Guide to the Common Weeds of Kansas*. Lawrence: University Press of Kansas, 1983.

Clark, Lynn G., and Richard W. Pohl. *Agnes Chase's First Book of Grasses*. 4th ed. Washington, D.C.: Smithsonian Institution Press, 1996.

Coffey, C. R., and R. L. Stevens. *Grasses of Southern Oklahoma and North Texas: A Pictorial Guide*. Ardmore, Okla.: Samuel Roberts Noble Foundation, 2004.

Costello, David F. *The Prairie World*. Minneapolis: University of Minnesota Press, 1980.

Couplan, Francois. *The Encyclopedia of Edible Plants of North America*. New Canaan, Conn.: Keats Publishing, 1998.

Cushman, Ruth C., and Stephen R. Jones. *The Shortgrass Prairie*. Boulder, Colo.: Pruett Publishing, 1988.

Darke, Richard. *The Encyclopedia of Grasses for Livable Landscapes*. Portland, Oreg.: Timber Press, 2007.

Editors of Time-Life Books. *Planet Earth: Grasslands and Tundra*. Alexandria, Va.: Time-Life Books, 1985.

Ely, Charles A., Marvin D. Schwilling, and Marvin E. Rolfs. *An Annotated List of the Butterflies of Kansas*. Hays, Kans.: Fort Hays State University, 1986.

Francaviglia, Richard V. *The Cast Iron Forest: A Natural and Cultural History of the North American Cross Timbers*. Austin: University of Texas Press, 2000.

Gates, Frank C. *Grasses in Kansas*. Topeka: Kansas State Printing Plant, 1936.

Gilmore, Melvin R. *Uses of Plants by the Indians of the Missouri River Region*. Lincoln: University of Nebraska Press, 1977.

Gould, F. W. *Grass Systematics*. New York: McGraw-Hill, 1968.

Great Plains Flora Association. *Flora of the Great Plains*. Lawrence: University Press of Kansas, 1986.

Haddock, Michael J. *Wildflowers and Grasses of Kansas*. Lawrence: University Press of Kansas, 2005.

Harrington, Harold D. *How to Identify Grasses and Grasslike Plants.* Athens, Ohio: Swallow Press, 1977.

Harris, James G., and Melinda W. Harris. *Plant Identification Terminology: An Illustrated Glossary.* 2nd ed. Spring Lake, Utah: Spring Lake Publishing, 2001.

Hitchcock, Albert S. *Manual of the Grasses of the United States.* 2nd ed. 2 vols. New York: Dover Publications, 1971.

Johnsgard, Paul A. *This Fragile Land: A Natural History of the Nebraska Sandhills.* Lincoln: University of Nebraska Press, 1995.

Jones, Stephen R. *The Last Prairie: A Sandhills Journal.* New York: Rugged Mountain Press, 2000.

Kartesz, J. T. North American Plant Atlas. Chapel Hill, N.C.: Biota of North America Program (BONAP), 2011, http://www.bonap.org/ MapSwitchboard.html.

Kaul, Robert B., David M. Sutherland, and Steven B. Rolfsmeier. *The Flora of Nebraska: Keys, Descriptions, and Distributional Maps of All Native and Introduced Species That Grow outside Cultivation.* Lincoln: Conservation and Survey Division, School of Natural Resources, Institute of Agriculture and Natural Resources, University of Nebraska, 2006.

Knoble, Edward. *Field Guide to the Grasses, Sedges and Rushes of the United States.* New York: Dover Publications, 1977.

Kucera, Clair L. *The Grasses of Missouri.* Columbia: University of Missouri Press, 1998.

Lady Bird Johnson Wildflower Center, http://www.wildflower.org/plants/.

Madson, John. *Where the Sky Began: Land of the Tallgrass Prairie.* Iowa City: University of Iowa Press, 2004.

Martin, Alexander C., Herbert S. Zim, and Arnold L. Nelson. *American Wildlife and Plants.* London: Constable, 1951.

Missouri Botanical Garden Grass Statistics, http://www.missouribotanical garden.org/.

Native Rangeland Grasses of the Texas Panhandle, http://www.northrolling plains.com/.

North American Grass Manual on the Web, http://herbarium.usu.edu/ webmanual/.

Pohl, Richard. *How to Know the Grasses.* Dubuque, Ia.: W. C. Brown, 1954.

Quayle, William A. *The Prairie and the Sea.* Cincinnati, Ohio: Jennings and Graham, 1905.

Raff, Marilyn. *Ornamental Grasses for Western Gardens*. Boulder, Colo.: Johnson Books, 2006.

Reichman, O. J. *Konza Prairie: A Tallgrass Natural History*. Lawrence: University Press of Kansas, 1987.

Savage, Candace. *Prairie: A Natural History*. Vancouver: David Suzuki and Greystone Books, 2004.

Shaw, Robert B. *Grasses of Colorado*. Boulder: University Press of Colorado, 2008.

Shirley, Shirley. *Restoring the Tallgrass Prairie*. Iowa City: University of Iowa Press, 1994.

Stubbendieck, James, Stephan L. Hatch, and Kathie J. Kjar. *North American Range Plants*. Lincoln: University of Nebraska Press, 1982.

Stubbendieck, James, and Kay L. Kottas. *Common Grasses of Nebraska*. Lincoln: University of Nebraska Extension, 2005.

Texas Invasive Plant Conference. Old World Bluestem Symposium. 2007, http://www.texasinvasives.org/professionals/converence07.php.

Tyrl, Ronald J., Terrence G. Bidwell, and Ronald E. Masters. *Field Guide to Oklahoma Plants: Commonly Encountered Prairie, Shrubland, and Forest Species*. Department of Plant and Soil Sciences. Stillwater: Oklahoma State University, 2002.

USDA, NRCS. The PLANTS Database. National Plant Data Team, Greensboro, NC 27401-4901 USA, http://plants.usda.gov, 24 July 2012.

Weaver, John E. *Native Vegetation of Nebraska*. Lincoln: University of Nebraska Press, 1965.

———. "North American Prairie" (1954.) Papers of John E. Weaver (1884–1956). Paper 15, http://digitalcommons.unl.edu/agronweaver/15/.

Whitson, Tom D. *Weeds of the West*. 9th ed. Darby, Pa.: Diane Publishing, 2006.

Index

Page numbers in bold type indicate the location of the main grass description, illustrations, and distribution map.